Lecture Notes in Economics and Mathematical Systems 523

Springer
Berlin
Heidelberg
New York
Hong Kong
London
Milan
Paris
Tokyo

Wei-Bin Zhang

A Theory of Interregional Dynamics

Models of Capital, Knowledge and Economic Structures

Springer

Author

Prof. Wei-Bin Zhang
Ritsumeikan Asia Pacific University
Jumonjibaru, Beppu-Shi
874-8577 Oita-ken
Japan

Cataloging-in-Publication Data applied for

A catalog record for this book is available from the Library of Congress.

Bibliographic information published by Die Deutsche Bibliothek
Die Deutsche Bibliothek lists this publication in the Deutsche Nationalbibliografie; detailed bibliographic data is available in the Internet at http://dnb.ddb.de

ISSN 0075-8450
ISBN 3-540-44306-1 Springer-Verlag Berlin Heidelberg New York

Springer-Verlag Berlin Heidelberg New York
a member of BertelsmannSpringer Science+Business Media GmbH

http://www.springer.de

Typesetting: Camera ready by author
Cover design: *Erich Kirchner*, Heidelberg

Printed on acid-free paper SPIN: 10894100 41/3142/du 5 4 3 2 1 0

Acknowledgements

I am grateful to the anonymous referees for valuable comments. I would like to thank Publishing Director Economics and Management Science Dr. Werner A. Müller and Editor Christiane Beisel for effective co-operation.

Chapters 2 to 11 are based on my published or unpublished manuscripts. Grateful acknowledgment is made to the following sources for the use of my published materials:

Chapter 2 for Zhang (1998a, 1998b);
Chapter 3 for Zhang, W.B. (1995a, 1998a);
Chapter 4 for Zhang, W.B. (1998b, 1995e);
Chapter 5 for Zhang, W.B. (1995b);
Chapter 6 for Zhang, W.B. (1993b);
Chapter 7 for Zhang, W.B. (1995c, 1995d);
Chapter 8 for Zhang, W.B. (1998a);
Chapter 9 for Zhang, W.B. (1997b);
Chapter 12 for Zhang, W.B. (1993c).

Preface

Over more than two centuries the development of economic theory has created a wide array of different theories, concepts and results. Nevertheless, there is no general theory, which unifies these varied theories into a comprehensive one. Economics has been split between partial and conflicting representations of the functioning of market economies. We have a collection of separate theories such as the Marxian economics, the Keynesian economics, the general equilibrium theory, and the neoclassical growth theory. These diverse economic theories have co-existed but not in a structured relationship with each other. Economic students are trained to understand economic phenomena by severally incompatible theories one by one in the same course. Since the end of Second World War many crises in economic theory have been announced. The economist experienced the crisis of the general equilibrium economics, the crisis of the neoclassical growth economics, the crisis of the Keynesian economics, not to mention the crises of the Marxian economics. It is quite reasonable to expect the loss of confidence in theoretical economics even among professional economists after so many crises in a very short period of time. But a crisis offers new opportunities for change, either for better or for worse. The past crises in theoretical economics may be perceived as a historical opportunity to construct a general economic theory by which the traditional theories are integrated into a higher whole.

In fact, my recent books, *Capital and Knowledge* (Zhang, 1999), *A Theory of International Trade* (Zhang, 2000), and *An Economic Theory of Cities* (Zhang, 2001), show how separate economic theories such as the Marxian economics, the Keynesian economics, the general equilibrium theory, the neoclassical growth theory, and the neoclassical trade theory, and the neoclassical urban economics can be examined within a single theoretical framework. This book is to further expand the framework proposed in the previous studies to an interregional context. The four books together demonstrate a comprehensive approach to economic complexity.

This book is a part of my economic theory with endogenous population, capital, knowledge, preferences, sexual division of labor and consumption, institutions, economic structures and exchange values over time and space (Zhang, 1996a). As an extension of the *Capital and Knowledge*, which is focused on the dynamics of national economies, this book is to construct a theory of interregional economies. We are concerned with dynamic relations between division of labor, division of consumption and determination of prices structure among regions. We examine dynamic interdependence between capital accumulation, knowledge creation and utilization, economic growth, price structures and interregional trade patterns under free competition. The theory is constructed on the basis of a few concepts within a compact framework. The comparative advantage of our theory is that in providing rich insights into complex of interregional economies it uses only a few concepts and simplified functional forms and accepts a few assumptions about behavior of consumers, producers, and institutional structures.

This book constructs a theoretical framework that would permit valid generalizations from one special modeling structure to another, and would deepen our understanding of economic evolution. It is a part of my broad approach to revealing complex of economic evolution (Zhang, 1991, 1996, 1999, 2000, 2001). I wish that the reader would appreciate this book within the framework that I have made great efforts to construct. The proper evaluation of the book should be done by perusing and digesting it as a part of the edifice, rather than by scanning it as a diamond.

Wei-Bin Zhang
Beppu-Shi, October 2002

Contents

1 Introduction

Economic activity of all types is moving in the direction of globalization. There is an increasingly expanding web of linkages and interconnections between states, societies and organization that make up contemporary world economic systems. Irrespective of the significance of understanding regional economic phenomena, regional economic theory is a relatively neglected area of economic analysis. Except many possible reasons for the negligence, it is argued that analytical difficulties associated with modeling dynamics of interregional economic interactions are a main obstacle to progress in regional economics. Isard (1956) assures: *"the general theory of location and space-economy is conceived as embracing the total spatial array of economic activities, with attention paid to the geographical distribution of inputs and outputs and the geographical variations in prices and costs".* The necessity of establishing dynamic regional theories has been recognized. Nevertheless, the traditional equilibrium economics and neoclassical growth theory, which have formed the two mainstreams of theoretical economics, fail to explain the complexity of economic geography. Although regional scientists and regional economists have proposed various kinds of regional dynamic models (e.g., Richardson, 1977, Henderson, 1985, Arnott, 1996a, 1996b, Zhang, 1997a, Fujita et al., 1999, Walz, 1999), as argued by Scotchmer and Thisse (1992), the current models do not integrate the spatial factor satisfactorily from the theoretical point of view. As they demonstrate in their comprehensive review on economics with space, "*integrating space and competition demands at the same time a certain dose of abnegation (abandon the familiar model of perfect competition) and of imagination (invent a new model of competition)*" (Scotchmer and Thisse, 1992). This book builds a few dynamic models of interregional economies to illustrate a way of *"embracing the total spatial array of economic activities [with an integration of space and competition, as well as] with attention paid to the geographical distribution of inputs and outputs and the geographical variations in prices and costs".*

In this book regional economics refers the study of economic phenomena that occur within and among regions. Our concern is limited to economic interactions among regions within the country but not among countries. In regional economics factors such as people and capital are free to move; while in international economics some factors (such as labor) may not be allowed to move beyond the national boundary. Differences in degrees of mobility of factors and goods within a country and between countries suggests that international trade theory may be invalid for addressing issues related to patterns of interregional trade and income distribution. At this initial stage, I distinguish international and regional economics mainly by whether or not people are free to move among the areas under consideration. The distinction between interregional and international economics might not be very strict; – factor mobility itself should be an endogenous variable of dynamic economics. My previous books, *A Theory of International Trade* (Zhang, 2000) focuses on trade issues between countries, and *An Economic Theory of Cities* (2001) is concerned with the complexity of cities, respectively. This book examines economic interactions among regions within the same analytical framework proposed in Zhang (1999). This book and the others just mentioned demonstrate a comprehensive approach to spatial economics.

Natural endowment differences in, for instance, climate and natural resources, lead to regional divergences in life style, consumption structure, and productivity. Regional factors such as climates, accessibility, infrastructures, land and natural resources, and regional cultures, affect economic geography in different ways. Issues arise due to the importance of non-traded goods (such as climate, land) and services (which are consumed simultaneously as they are produced) in analyzing the movements of product factors, such as labor and capital within a country. For instance, mobile labor consumes in the location where it works. Workers will move to the location at which they can achieve higher utility. The factor mobility will equalize utility levels across regions, but not necessarily wage rates. Wages tend to be higher in locations where the cost of nontraded goods is higher to compensate workers for the higher cost of living. It is significant to examine how these endowments affect wages, rents, and service prices. From studies on regional economies with regional differences of wages and land rent, it is discerned that it is difficult to analyze spatial economic phenomena with natural endowment differences even when we purposefully limit the investigation to static economies.

Industrialization, urbanization, population growth and migration drive economic geography to materialize differently over time. In modern times, global as well as national economic geographies have dramatically changed in association with sectoral adjustments from agriculture to manufacturing and from manufacturing to services. Economic growth of multiregional systems is one of the main issues in regional economics and regional sciences. Numerous models of regional growth have been proposed to deal with various aspects of regional economic structures and growth.

The neoclassical economic growth theory characterized by the Solow-Swan model had an important influence on the development of regional growth theory. It is shown that a two-region neoclassical growth model under certain conditions has an equilibrium at which regional growth rates are equalized (Borts and Stein, 1964, Siebert, 1969). The neoclassical regional growth theory is based on the Solow-Swan's tradition. It is well known that the long-term growth rate in the Solow-Swan model is exogenously determined. In particular, the steady-state growth rate of per capita income is equal to the rate of exogenous technological change or population growth rate in these models. In order to allow for different regional growth paths, differences in the rate of regional technological change should be assumed. The neoclassical growth theory assumes that the market mechanism works so perfectly that any disparity in regional wages tends to disappear. Differences in productivity growth are considered as being fundamentally due to interregional misallocation of resources. Any disparities in productivity growth will eventually vanish as the misallocation is progressively corrected and regional growth rates approach the steady state.

Some economists were cognizant of the causes of spatial differences in long-run growth rates and the causes of productivity improvements such as innovation, economies of scales, and learning-by-doing. But these later causes violate the assumptions of perfect assumptions and constant returns to scale; it is difficult to formally take these causes into account when modeling regional growth. There are some contemporary studies of increasing returns; most of those works deal with economies of increasing returns under imperfect competition.

Agglomeration economies are associated with increasing returns from specification, from the higher rate of technological change and spillover efforts, and from the higher rate of human capital accumulation. Myrdal (1957) brings about ideas of the "circular and cumulative causation" to examine multiregional growth. The cumulative causation theory treats increasing returns to scale as the main source of regional growth. Kaldor (1966) argues how export led regional growth impulses may set off self-reinforcing growth processes through agglomeration economies and the achievement of successively higher market thresholds. It is held that linkages between growing and lagging regions may not be equally beneficial. Dixon and Thirlwall (1975) formally model Kaldor's ideas about regional development.

It is observed that there are substantial differences in labor productivity among the manufacturing sectors of the various regions of the United States. Moomaw (1983) reviews the literature on the study of existence of spatial differentials in labor productivity and possible implications of these differentials for regional and urban development. Another early study carried by Verdoorn demonstrates that manufacturing industry is subject to substantial economies of scale (McCombie, 1988). It is argued that faster growing regions will experience a greater growth of productivity than the more slowly growing regions. Rapidly growing regional competitiveness is increased in the manner of cumulative causation. McCombie (1988) examined the causes of regional disparities in growth and unemployment within the neoclassical growth framework. One of the causes of disparities in regional productivity growth is the eradication of an initial interregional misallocation of resources. McCombie also analyzed the impact of a progressive reduction in the degree of intraregional misallocation of resources on productivity growth and emphasized the role of the spatial diffusion of innovations in regional productivity.

In examining agglomeration economies, Krugman (1989, 1991) develops a few models of economic geography. He tries to incorporate regional aspects such as factor mobility, transport costs, and the interaction of agglomeration advantages and costs. Abel-Rahman (1988) explains urban agglomeration and regional equilibrium on the basis of the Dixit-Stiglitz approach of monopolistic competition to determine equilibrium city size. The model uses the concept of product differentiation as an important factor in the formation of large cities. The supply side is characterized by monopolistic competition and decreasing average cost at the firm level. For the demand side, product variety is presented as a key factor in consumer agglomeration. Consumers buy differentiated products available in the local market. Consumer agglomeration in a particular area enlarges the equilibrium number of firms selling differentiated products, which raises utility in the Dixit-Stiglitz manner and at the same time augments the city's radius and raises rents at any given location. The equilibrium is achieved by the interactive forces of product diversity gains, consumer agglomeration and changes in transport costs and rents. Abel-Rahman shows that in a system of two cities with no trade, at equilibrium all workers may not concentrate in the city with more product variety as is argued by Krugman (1979). This shows the importance of spatial consideration in regional trade models. Applying the frameworks proposed by Ethier (1982) in trade theory and Romer (1987) in growth theory, Rivera-Batiz (1988) proposes a model that endogenizes agglomeration economies from both the production and consumption sides. It is assumed that an increase in city sizes enlarges the variety of consumer services locally available, shifting upwards household utility, and is associated with an agglomeration of industrial producers that raises the derived demand for local business, allowing increased specialization among them. The increased number of producer or business services available in the city improves the productivity of the industrial base of that city and results in endogenous external economies of scales.

Modern developed economies are knowledge-based in the sense that knowledge creation and utilization play the key role for rise and decline of national economies. In knowledge-based economies, knowledge creation and utilization also play the key role for the rise or decline of regions. Changes in technology and productivity are the key variables in explaining regional economic phenomena. Although natural resources, regional amenity and infrastructures are not highly mobile, the population and the skills and available human capital associated with people are mobile and flexible. It has been observed that regions that become important loci of invention stand to gain enormous economic and political influence (Suarez-Villa, 1993). The development of major loci of invention tends to be associated with high qualitative level of human capital resources and infrastructure with long-lasting effects on industry, services, competitiveness, and migration. The development of an innovative region is usually accompanied by the creation of radically new economic activities, markets, and technological applications. Highly skilled individuals and labor tend to be attracted to regions with good inventive infrastructures.

The neo-Ricardian notation of land-locked 'comparative advantage' is not sufficient to explain phenomena associated with 'constructed advantages' with knowledge (Varga, 1998). The approach of spatial economy by Nelson and Winter (1982) dispense with the neoclassical conventions of continuous production and exogenous technological progresses. In this approach the technological progress occurs from processes of economic search and selection. Economic search occurs when firms are induced by declining profits to seek new technologies. Firms which succeed in finding efficient technologies will expand; while these which are not so successful will contract. The processes of search and selection together consist of the economic sources of increasing technological efficiency of the aggregate economy through time.

Influenced by the recent works in international trade theory, Englmann and Walz (1995) propose a spatial model of economic growth with endogenous technological change in the two regions. They examine the determinants of regions' growth rates in the presence of factor mobility. In this approach, regional specification patterns arise from the interaction of transport costs in combination with economies of scale and a given regional endowment with an immobile factor. They analyze the long-run effects of a variety of regional policy instruments. They show that under certain circumstances various policies would not achieve their intended targets. For instance, a regional policy which provides subsidies to traditional-goods producers and is supposed to reduce regional gaps in income actually enlarge the gap due to returns to scale economies. A political instrument of making investment in interregional transport infrastructure in the less-advanced region may reduce the region's location advantage either by decreasing the available number of immobile workers for production in the industrial sector or by increasing price competition in the less-advanced region. The impact of certain regional policies are situation-dependent. The regional policy with a good will may turn to be the opposite. This also implies that regional policy makers should be fully aware of each region's advantages and disadvantages and each region's special situations in order to be able to implement sensible policies in an effective way. Simply providing the less-developed region good conditions may do not have the positive impact in the presence of mobile factors of productions, footloose industries and non-constant returns to scales.

A key difference between the spatial models influenced by the new growth theory and new trade theory and my approach is that I take capital and knowledge into a single consistent framework, while most of the spatial models with endogenous knowledge analyze regional growth and technological progress, omitting physical capital. The remainder of the book is organized as follows.

Chapter 2 proposes a compact two-region economic model with endogenous capital accumulation. Each region is characterized by its preference structure and its immobile resources such as regional amenity and land. The model describes an interaction between capital accumulation, regional distribution of capital and labor, and locational prices over time under perfectly competitive institution. Section 2.1 defines the basic model. Section 2.2 proves the existence of a unique equilibrium of the dynamic system when the two regions have identical propensities to consume goods, to consume lot size and to own wealth. Sections 2.3 and 2.4 examine respectively the effects of changes in a region's working efficiency and amenity level on the regional economic equilibrium structure. Section 2.5 concludes the chapter. In the appendix, we examine the conditions for existence of economic equilibria when the two regions have different propensities to consume goods, to consume lot size and to own wealth.

Chapter 3 provides a compact two-region economic model with endogenous time distribution, consumption and capital accumulation. Each region is characterized by its preference structure and its immobile resources such as regional amenity and land. The model describes an interaction between capital accumulation, regional distribution of capital and labor, time distribution of each worker in each region, and locational prices over time under perfectly competitive institution. We provide the conditions for the existence of equilibria. We analyze how changes in regional amenity and working efficiency may affect the economic equilibrium structure.

Chapter 4 builds a compact two-region economic model with sexual division of labor, endogenous time distribution, consumption and capital accumulation. Each region is characterized by its preference structure and its immobile resources such as regional amenity and land. The model describes an interaction between capital accumulation, sexual division of labor, regional distribution of capital, time distribution of each worker in each region, and spatial prices over time under perfectly competitive institution. We provide the conditions for existence of economic equilibria and analyze how changes in some parameters may affect the equilibrium structure.

Chapter 5 sets up a compact two-group and two-region economic model with endogenous capital accumulation. The economic system consists of two groups of people and two regions. Each group is characterized by its preference and working efficiency and each region is characterized by its preference structure and its immobile resources such as regional amenity and land. The model describes an interaction between capital accumulation, regional distribution of capital and labor, and locational prices over time under perfectly competitive institution. We provide the conditions for the existence of equilibria. We analyze how changes in regional amenity and working efficiency may affect the economic equilibrium structure.

Chapter 6 devises an analytical framework within which the basic ideas of the two approaches by von Thünen and the Alonso model are combined. We contrive an equilibrium model of economic geography with two cities and one farm. The model provides equilibrium conditions of a perfectly competitive spatial economy. The economic system consists of one agricultural and two industrial sectors. It is assumed that all the households have an identical preference and that any possible costs for professional changes and people movement among cities and countryside are omitted. The cities and the countryside are different in amenity and technology. This chapter examines how division of labor and difference in amenity among the urban and rural areas may affect spatial distribution of economic variables. The chapter is organized as follows. Section 6.1 defines the two-city and one farm economic geographic model. Section 6.2 provides the conditions for existence of equilibria. Section 6.3 examines the effects of changes in city 1's amenity on the economic structure. Section 6.4 concludes the chapter. The appendix proves the proposition in Section 6.2.

Chapter 7 proposes an equilibrium model of economic geography with two cities and one farm. It has now become clear that there are a number of potentially important spatial influences, such as public goods, amenities, different externalities, transportation costs, that may challenge the validity of competitive equilibrium theory for explaining a regionally heterogeneous economy. For instance, one difficult factor is the so-called capitalization, which implies that the price of land is interdependent with local amenities, economic agents' densities, transportation costs and other local variables or parameters. Although the significance of capitalization has been noticed by location theorists, it may be said that we still have no compact framework within which we can satisfactorily explain the issue. The purpose of this chapter is to discuss the issue within a perfectly competitive equilibrium framework. The main contribution of this chapter is to generate a two-region economy within an equilibrium framework with constant return-to-scale economies and regional immobile factors of land and amenity. This chapter is organized as follows. Section 7.1 defines the two-region equilibrium model. Section 7.2 provides equilibrium conditions. Section 7.3 examines the effects of changes in the population of the farmer class on the economic structure. Section 7.4 analyzes the impact of changes in region 1's amenity for the landlord class on the economic structure. Section 7.5 concludes the chapter.

Chapter 8 constructs a two-region and two-class equilibrium model of perfect competition. In the past two decades, analysts have become increasingly aware of the need to improve the specification of household behavior in the input-output models. This chapter proposes a new framework for examining structural issues within an interregional context. To make an initial attempt to synthesize ideas in location theory, urban economics, economic geography, equilibrium theory and neoclassical growth theory, we construct a dynamic model to describe the interaction between capital accumulation, the regional distribution of capital and labor, the division of labor, the capital distribution in each region, land rents, service prices and commodity prices. The model includes three spatial factors: land, amenities and services. A regional analysis must treat services differently from commodity goods since many services provided by schools, hospitals and restaurants must be consumed where they are supplied. The spatial character of services implies that service prices will be spatially varied. Amenities such as climatic conditions and historical buildings are location-dependent and have an important impact upon residential location. The chapter organized as follows. Section 8.1 defines the basic model. Section 8.2 analyzes properties of the dynamic system. Section 8.3 examines effects of changes in the amenity level on the equilibrium structure of economic geography. Section 8.4 concludes the chapter.

The objective of Chapter 9 is to study relationships between regional growth and regional trade patterns in the context of perfectly competitive institution. The national economy composes two regions. Each country produces goods and makes contribution to knowledge growth. Each region's production is similar to the standard one-sector growth model. Knowledge accumulation is through learning by doing. The regional trade aspects of our model is similar to the international macroeconomic one-sector growth model with perfect capital mobility. It is well known that many efforts have been made to examine the impact of savings, technology and various policies on the trade patterns within the framework of dynamic models with capital accumulation. Our modeling framework is based on the dynamic one-sector and two-country framework. We are interested in how difference in knowledge utilization efficiency between two regions may affect pattern of regional trade and national economic development. The main differences between the models just mentioned and this chapter are related to labor mobility. This chapter assumes free mobility of labor force between regions. This chapter is organized as follows. Section 9.1 defines the basic model. Section 9.2 provides the conditions for existence of equilibria of the dynamic system. Section 9.3 examines the effects of changes in each region's amenity level, the population and each region 1 's knowledge accumulation efficiency on the equilibrium structure of the regional economy. Section 9.4 concludes the chapter. In Appendixes A.9.1 and A.9.2, we study the impact of changes in the population and each region's knowledge accumulation efficiency on the equilibrium structure.

Irrespective of many efforts on international trade, there were only a few formal economic models which deal with interdependence between interregional economic growth and knowledge accumulation. Many interregional aspects remain to be further examined. Chapter 10 combines the main ideas of Chapters 8 and 9. We propose a two-region and multi-sector model of perfect competition with endogenous knowledge to examine the role of knowledge in the interregional division of labor and consumption. In addition to explicitly modeling endogenous knowledge, we also take account of some other important aspects of economic geography. We emphasize the geographical character of services. This chapter is organized as follows. Section 10.1 defines a two-region and multi-sector economic model with endogenous knowledge accumulation. Section 10.2 shows that a dynamic system may have either a unique or multiple equilibria and each equilibrium may be either stable or unstable, depending upon knowledge utilization and creativity of the different sectors. Section 10.3 examines the effects of changes in the efficiency of knowledge accumulation and amenities upon the equilibrium economic geography. Section 10.4 concludes the study. The appendix offers the proof of the results presented in Section 10.2.

Chapter 11 is to introduce education into the interregional modelling framework proposed in the preceding chapters. It is important to examine the role of the university education and research on economic development at the regional scale. Universities play a central role in regional innovation systems as producers of basic research. The importance of basic research is due to the public good nature of the research and positive externalities to the private sector in the form of knowledge spillovers. Regional economies have experienced dramatic changes in different parts of the world in the last few decades. These changes in economic development are influenced by increases of the share of the total resources devoted to R&D, in the educational achievements of the labor force and in the rate of technological change. This chapter proposes a dynamic endogenous two-region growth model. One of the regions is assumed to contain a research university - the only university in the whole economy. The chapter is organized as follows. The basic model is presented in Section 11.1. In Section 11.2, we analyze the conditions for the existence of equilibria. The impacts of policies of high education are analyzed in Section 11.3; the impacts of changes in regional amenities are analyzed in Section 11.4. Section 11.5 concludes the chapter.

Chapter 12 introduces endogenous knowledge to location theory to examine how economic spatial structure may be affected by knowledge. For simplicity of analysis, we are concerned with dynamics of an isolated economic system consisting of three - agricultural, industrial and service - kinds of economic activities. The geography of our economic system is composed of the three parts, i.e., the CBD, the residential area and the agricultural area. Different economic activities are located spatially differently. We assume that industrial production and service supply are located at the CBD, urban households are located in the residential area (as in the Alonso model), and farming and farmers' dwelling location are distributed over the rural area. The model determines spatial and temporal distribution of real variables and monetary variables under the assumption of perfect competition in all the markets.

2 Regional Growth with Productivity and Amenity Differentials

Regions are different in living conditions and productivity. It is observed that tropical people tended to spend less time and energies on working and thinking than people living in regions with proper climates. Regional differentials in natural conditions and cultural attitudes towards work and knowledge play a substantial role in determining regional differences in economic evolution. Nevertheless, economics has not provided a consistent framework to analyze the role of regional conditions such as natural resources and cultural attitudes (toward playing, working, saving, learning, creativity) on economic evolution. This chapter proposes a model of regional economic development based on the growth theory proposed in Zhang (1999).

Firms and households take regional differentials in living conditions and productivity into account in their locational decision-making. The productivity advantages of one region may be offset to some extent by the higher wages that must be paid in a system where people are free to choose where they work and live. Higher wages are often associated with some kinds of disamenities (such as noise, pollutants, and densely populated neighborhood) and high living costs. Labor force, capital, knowledge are main inputs of economic production. Labor and capital are easily mobile between regions in industrialized economies. As capital mobility becomes high and costs associated with capital movement among regions becomes low, it is reasonable to assume that capital movement equalizes rate of interest between regions within a national economy. But there are different principles for analyzing temporary equilibrium conditions for labor movement in a dynamic regional framework. For example, in the theoretical literature on regional economics two principles have been proposed to analyze labor movement. One is that labor movement, if costs associated with professional or locational changes are neglected, equalizes wage rates between regions. This assumption is limited in a freely competitive national labor market if various regions provide different levels of amenity (such as regional cultures, climates and pollution) and have different technologies. The other one is that free movement of people equalizes utility levels, which they may obtain in different regions. Although this assumption is reasonable for analyzing issues related to movement of people between regions, it is not easy to apply the principle within a compact dynamic framework with regional differences in life styles and wage rates. A reason is that it is difficult to examine issues related to ownership of resources and capital in a dynamic regional economy. If wage rates and life style (including saving behavior) are regionally different and people may freely move within the economy, it can be seen that how to calculate each individual's capital and his income in a dynamic multi-regional economy is not an easy matter. In this chapter, we solve this issue by utilizing a utility function recently used by Zhang (e.g., 1999). In this approach we assume that each household's preference structure over lot size, consumption and wealth is represented by the utility function of the region where it chooses to live and work at any point of time t. Wealth may practically be accumulated for different reasons such as the capitalist spirit, old age consumption, providing education for children, power and social status. It should be remarked that in our approach a "far-sighted" household's preference includes wealth, even though he may not get "direct utility" from holding wealth, for instance, as a symbol of status or success. The far-sighted household's preference for consumption flows over the future is manifested in his preference for wealth. This preference for wealth is manifested in his current utility function at each point of time.

This chapter proposes a simple compact two-region growth model of endogenous capital accumulation under perfect competition and free movement of the population with given regional differences in land, amenity and life styles. Section 2.1 defines the basic model. Section 2.2 proves the existence of a unique equilibrium of the dynamic system when the two regions have identical propensities to consume goods, to consume lot size and to own wealth. Sections 2.3 and 2.4 examine respectively the effects of changes in a region's working efficiency and amenity level on the regional economic equilibrium structure. Section 2.5 concludes the chapter. In the appendix, we examine the conditions for existence of economic equilibria when the two regions have different propensities to consume goods, to consume lot size and to own wealth. This chapter is based on Zhang (1998a, 1998b).

2.1 Regional Growth with Differences

We consider an economic system consisting of two regions, indexed by 1 and 2, respectively. Each region has a single production sector. Production is carried out by combining capital and labor. It is assumed that the two regions' product is qualitatively homogenous and is either consumed or invested. We assume a homogenous population. A person is free to choose his residential location. We assume that any person chooses the same region where he works and lives. Each region has fixed land. Land quality, climates, and environment are homogenous within each region, but they may vary between the two regions. The land is used only for housing. We select commodity to serve as numeraire, with all the other prices being measured relative to its price. We neglect transportation cost of commodities between and within regions. As become evident later on, although it is conceptually not difficult to introduce transportation cost function and to provide balance conditions for demand and supply and for price equalization conditions with transportation cost (e.g., Takayama and Judge, 1971, Roy and Johansson, 1993), the problem will become analytically too complicated. The assumption of zero transportation cost of commodities implies price equality for the commodity between the two regions. As amenity and land are immobile, wage rates and land rent may be not equal between the two regions.

We introduce

N — the given population of the economy;
L_j — the given territory size of region j, $j = 1, 2$;
$K(t)$ — the total capital stocks of the economy at time t;
$F_j(t)$ — the output levels of region j 's industrial sector at time t;
$K_j(t)$ and $N_j(t)$ — the levels of capital stocks and labor force employed by region j 's production sector;
$c_j(t)$ and $s_j(t)$—per capita's consumption level of commodity and savings made by per capita in region j;
$l_j(t)$—the lot size per capita in region j;
$y_j(t)$—the net income per capita in region j;
$r(t)$ — the rate of interest; and
$w_j(t)$ and $R_j(t)$—region j 's wage rate and land rent, respectively.

We now describe the model.

The production sectors
The production functions of the two regions are specified as follows

$$F_j = K_j^{\alpha_j}(z_j N_j)^{\beta_j}, \quad \alpha_j + \beta_j = 1, \quad \alpha_j, \beta_j > 0, \quad j = 1, 2 \quad (2.1.1)$$

in which z_j are the working efficiency index of region j 's labor force. For simplicity, we assume z_j to be constant. From the literature of human capital and economic growth, it is conceptually not difficult to introduce dynamics of z_j. The marginal conditions are given by

$$r = \frac{\alpha_j F_j}{K_j}, \quad w_j = \frac{\beta_j F_j}{N_j}, \quad j = 1, 2.$$

The rate of interest is identical in the whole economy and the wage rates w_j may be different between the two regions.

Behavior of households

Each worker may get income from land ownership, capital ownership and wages. In order to define net incomes, it is necessary to determine land ownership structure. It can be seen that land properties may be distributed in multiple ways under various institutions. To simplify the discussion, we specify land revenue as follows. We assume that each worker owns L_1 / N amount of land in region 1 and L_2 / N in region 2 and it is impossible to sell land but it is free to rent one's own land to others. Under this assumption, the land revenue $\bar{r}$ per worker is given as follows

$$\bar{r} = \frac{L_1 R_1 + L_2 R_2}{N}. \tag{2.1.3}$$

This assumption implies that a worker receives the land revenue $\bar{r}$ irrespective of whether his dwelling site is located in region 1 or region 2.

If we denote $k_j(t)$ capital stocks owned by per capita in region j, the interest payment per capita is given by $rk_j(t)$. Under the above specified land ownership distribution, the net income per capita y_j in region j consists of the wage income w_j, land revenue $\bar{r}$, and interest payment rk_j, i.e.

$$y_j = \bar{r} + w_j + rk_j. \tag{2.1.4}$$

It is assumed that there is an aggregated utility function for each region. It is assumed that a typical person's utility level, $U_j(t)$ in region j is dependent on the person's consumption level $c_j(t)$ of community, and the level of the person's 'gross savings', $s_j(t)$. For simplicity, the utility functions are specified as follows

$$U_j(t) = A_j l_j^{\eta_j} c_j^{\xi_j} \left(k_j + s_j - \delta_k k_j\right)^{\lambda_j},$$
$$\eta_j, \xi_j, \lambda_j > 0, \ \eta_j + \xi_j + \lambda_j = 1 \tag{2.1.5}$$

in which η_j, ξ_j, and λ_j are respectively a typical person's elasticity of utility with regard to lot size, commodity and savings in region j. We call η_j, ξ_j, and λ_j region j's propensities to consume lot size, to consume goods, and to save (or to hold wealth), respectively. We assume a homogeneous population in the national economy; but we consider that two regions may have different forms of utility functions. Consumer behavior is affected by many factors such as climates and regional cultures. People may have higher propensity to consuming housing in cold region than in warm one. For instance, a consumer may have different consumption patterns in different regions with the same budget. We take account of this difference in regional consumption by assuming possible differences in the two regions' utility functions.

In (2.1.5), A_j is called region j 's amenity level. Amenities in this chapter are considered as locational parameters or slow changing variables, such as infrastructures, regional cultures and climates, which may affect attractiveness of a location (e.g., Kanemoto, 1980, Diamond and Tolley, 1981, Blomquist, Berger and Hoehn, 1988). Amenities such as noise and pollution should be treated as fast variables. Here, we neglect possible issues related to speed of changes of amenity levels.

The form of utility functions in (2.1.5) is used in studies on economic growth with endogenous saving by Zhang (1999). In this approach we assume that each household's preference structure over lot size, consumption and gross savings is represented by the utility function of the region where it chooses to live and work at any point of time t. We assume that an increase in gross savings tends to increase utility level of a typical household. Decision on savings may practically be made for different reasons such as the capitalist spirit, old age consumption, providing education for children, power and social status. Those different reasons determine preference structure. In our approach a "far-sighted" household's preference includes savings, even though he may not get "direct utility" from holding wealth, for instance, as a symbol of status or success. The far-sighted household's preference for consumption flows over the future is manifested in his preference for savings. This preference for savings is manifested in his current utility function at each point of time. In other words, the above utility function also takes account of the possibility that households have incentives to engage in wealth accumulation quite aside from any direct utility they might obtain from holding wealth.

It may be necessary to remark that in a traditional intertemporal framework, the economy maximizes

$$\int_t^{\infty} U(C) \exp(-\rho t)\, dt$$

subject to the dynamic budget constraint of capital accumulation. In the above formula, there are two strict assumptions. The first is that utility is additional over time. Although we may add capital over time, it is a very strict requirement to add utility over time. It is not reasonable to add happiness over time. The second is that the parameter ρ is meaningless if utility is not additional over time. It is obvious that our formula does not involve these two issues. In our approach, we take account of social and cultural factors which affect saving behavior by the preference parameters at each point of time.

Each consumer j gets income y_j. He can also sell his property k_j to purchase consumption goods and make investment. The 'total available budget' for savings and consumption is equal to

$$y_j^* = y_j + k_j .$$

We assume that consumer j pays the depreciation of capital goods which he owns. The amount is equal to $\delta_k k_j$ where δ_k is the depreciation rate of physical capital. At each point of time consumer j would distribute among savings (s_j), consumption of goods (c_j), consumption of housing ($R_j l_j$) and payment for depreciation ($\delta_k k_j$). The budget constraints are thus given by

$$R_j l_j + c_j + \delta_k k_j + s_j = y_j^*, \quad j = 1, 2 . \tag{2.1.6}$$

A typical person maximizes the person's utility subject to its budget constraint. The optimal problems have the following solutions

$$R_j l_j = \eta_j \Omega_j, \; c_j = \xi_j \Omega_j, \quad s_j = \lambda_j \Omega_j - \delta k_j \tag{2.1.7}$$

where

$$\Omega_j \equiv y_j + \delta k_j, \quad \delta \equiv 1 - \delta .$$

According to the definitions of k_j and s_j, the capital accumulation of a typical person in region j is given by

$$\frac{dk_j}{dt} = s_j - \delta_k k_j .$$

Substituting s_j in (2.1.7) into these two equations yields

$$\frac{dk_j}{dt} = \lambda_j \Omega_j - k_j . \tag{2.1.8}$$

As households are assumed to be freely mobile between the two regions, consumers should obtain the same level of utility at any point of time, irrespective of in which region they live, i.e.

$$U_1(t) = U_2(t). \tag{2.1.9}$$

Balance of demand and supply

By the definitions of K, k_j and N_j, we have

$$K = k_1 N_1 + k_2 N_2. \tag{2.1.10}$$

The assumption that labor force, capital stocks and land are fully employed is represented by

$$N_1 + N_2 = N, \quad K_1 + K_2 = K, \quad l_j N_j = L_j, \quad j = 1, 2. \tag{2.1.11}$$

The balance of demand for and supply of industrial commodity is given by

$$(c_1 + s_1)N_1 + (c_2 + s_2)N_2 = F_1 + F_2. \tag{2.1.12}$$

We have thus built the model. The system has 25 variables, N_j, K_j, F_j, c_j, s_j, k_j, l_j, y_j, U_j, w_j, R_j, $(j = 1, 2)$, K, r and $\bar{r}$. It can be shown that the system contains the same number of independent equations. We now examine conditions for the existence of equilibria of the dynamic system.

2.2 Equal Propensities to Consume and to Hold Wealth

This section is concerned with conditions for the existence of equilibria of the dynamic system. For simplicity of analysis, we assume that the propensities to consume lot size, to consume goods, and to own wealth are spatially identical. In the appendix, we examine the case that the propensities are not identical between the two regions. As shown in the appendix, it is quite difficult to explicitly interpret the conditions for existence of economic equilibria when the two regions have different propensities.

Assumption 2.2.1.

In this section, we assume that the propensities are equal between the two regions, i.e., $\eta \equiv \eta_1 = \eta_2$, $\xi \equiv \xi_1 = \xi_2$, and $\lambda \equiv \lambda_1 = \lambda_2$.

We now examine conditions for existence of equilibria. By (2.1.8), at equilibrium we have

$$\lambda\Omega_j = k_j, \quad j = 1, 2. \tag{2.2.1}$$

It should be remarked that we might extend our investigation to the case of stationary growth by replacing (2.2.1) with $\lambda\Omega_j = (1 + g_j)k_j$ where g_j is region j's fixed growth rate.

Substituting (2.2.1) into (2.1.7) yields

$$l_j = \frac{\eta k_j}{\lambda R_j}, \; c_j = \frac{\xi k_j}{\lambda}, \quad s_j = \delta_k k_j. \tag{2.2.2}$$

By $l_j = \eta k_j / \lambda R_j$ in (2.2.2) and $l_j N_j = L_j$ in (2.1.11)

$$R_j = \frac{\eta k_j N_j}{\lambda L_j}. \tag{2.2.3}$$

Substituting c_j and s_j in (2.2.2) and $l_j N_j = L_j$ in (2.1.11) into (2.1.5) and then using (2.1.9), we get

$$\frac{k_1}{k_2} = A\bar{N}^{\theta\eta} \tag{2.2.4}$$

in which

$$A \equiv \left(\frac{A_2}{A_1}\right)^{\theta}\left(\frac{L_2}{L_1}\right)^{\theta\eta}, \quad \bar{N} \equiv \frac{N_1}{N_2}, \quad \theta \equiv \frac{1}{\xi + \lambda}.$$

By (2.2.2), (2.1.12) and (2.1.10), we get

$$\delta_1 K = F_1 + F_2 \tag{2.2.5}$$

where $\delta_1 \equiv \xi / \lambda + \delta_k$. By (2.1.3), (2.2.3) and (2.1.10), we get

$$\bar{r} = nK \tag{2.2.6}$$

where $n \equiv \eta / \lambda N$. By (2.2.1), (2.1.4) and the definitions of Ω_j, we get

$$w_j + nK = (a - r)k_j, \quad j = 1, 2 \tag{2.2.7}$$

where $a \equiv 1/\lambda - \delta > 0$. By (2.1.1) and $r = \alpha_1 F_1 / K_1 = \alpha_2 F_2 / K_2$ in (2.1.2), we solve

$$K_1 = \left(\frac{\alpha_1}{\alpha_2}\right)^{1/\beta_1} \left(\frac{K_2}{z_2 N_2}\right)^{\beta_2/\beta_1} z_1 N_1. \tag{2.2.8}$$

By (2.2.8) and $K_1 + K_2 = K$, we get

$$K_2 + \left(\frac{\alpha_1}{\alpha_2}\right)^{1/\beta_1} \left(\frac{K_2}{z_2 N_2}\right)^{\beta_2/\beta_1} z_1 N_1 - K = 0. \tag{2.2.9}$$

We can check that for given $N_j > 0$ and $K > 0$, the above equation has a unique positive solution

$$K_2 = f(N_1, N_2, K), \quad 0 < K_2 < K$$

where f is a function of N_1, N_2 and K. We thus can represent K_1 and K_2 as functions of N_1, N_2 and K. In particular, in the case of $\alpha_1 = \alpha_2$, by (2.2.9) and $K_1 + K_2 = K$, we have

$$K_1 = \frac{z\overline{N}K}{1 + z\overline{N}}, \quad K_2 = \frac{K}{1 + z\overline{N}} \tag{2.2.10}$$

where $z \equiv z_1 / z_2$.

Assumption 2.2.2.
We assume: $\alpha \equiv \alpha_1 = \alpha_2$ (or $\beta \equiv \beta_1 = \beta_2$).

This assumption implies that the elasticity of production with regard to capital are equal in the two regions. Regional production difference is difference in working efficiency parameters, z_1 and z_2. Under Assumption 2.2.1, we solve K_1 and K_2 as functions of N_1, N_2 and K in simple forms given in (2.2.10). As shown below, this assumption is accepted for simplicity of analysis. The case of $\alpha_1 \neq \alpha_2$ can be similarly discussed.

By (2.2.10), $K_1 / K_2 = z\overline{N}$ holds. From this equation and $w_j = \beta F_j / N_j = \beta z_j (K_j / z_j N_j)^{\alpha}$, we get

$$\frac{w_1}{w_2} = z, \quad \frac{F_1}{F_2} = z\overline{N}.$$

By (2.2.5), $F_1 / F_2 = z\overline{N}$, $r = \alpha F_1 / K_1$, and $w_j = \beta F_j / N_j$, we solve

$$F_1 = \frac{\delta_1 z\overline{N}K}{1 + z\overline{N}}, \quad K_2 = \frac{\delta_1 K}{1 + z\overline{N}}, \quad r = \alpha\delta_1, \quad w_1 = \frac{\beta\delta_1 z\overline{N}K}{(1 + z\overline{N})N_1},$$
$$w_2 = \frac{\beta\delta_1 K}{(1 + z\overline{N})N_2}. \qquad (2.2.11)$$

By $N_1 + N_2 = N$ and $\overline{N} = N_1 / N_2$, we have

$$N_1 = \frac{N\overline{N}}{1 + \overline{N}}, \quad N_2 = \frac{N}{1 + \overline{N}}. \qquad (2.2.12)$$

Substituting w_j in (2.2.11) into (2.2.7) and then dividing the two equations, we get

$$\Phi(\overline{N}) \equiv \frac{\beta\delta_1 z(1 + \overline{N}) + nN(1 + z\overline{N})}{\beta\delta_1(1 + \overline{N}) + nN(1 + z\overline{N})} - A\overline{N}^{\eta\theta} = 0, \quad 0 \leq \overline{N} < \infty$$
$$(2.2.13)$$

where we use (2.2.4) and (2.2.12). It can be shown that the function $\Phi(\overline{N})$ has the following properties

$$\Phi(0) > 0, \quad \Phi(+\infty) < 0, \quad \frac{d\Phi}{d\overline{N}} = -\eta\theta A\overline{N}^{\eta\theta-1}$$
$$- \frac{(z-1)^2 \beta\delta_1 nN}{\{\beta\delta_1(1 + \overline{N}) + nN(1 + z\overline{N})\}^2} < 0. \qquad (2.2.14)$$

By (2.2.14), we conclude that the equation, $\Phi(\overline{N}) = 0$, has a unique positive solution $\overline{N}$. We see that $\overline{N}$ is determined by (2.2.13) and N_j by (2.2.12). From (2.2.10), (2.1.1) and (2.2.5), we solve K as follows

$$K = \frac{N^*}{\delta_1^{1/\beta}} \tag{2.2.15}$$

where $N^* = z_1 N_1 + z_2 N_2$.

By (2.2.11), we have $r = \alpha \delta_1$. It is direct to check: $a - r > 0$. By (2.2.7), we see that $k_j > 0$. Summarizing the above discussion, we obtain the following lemma.

Lemma 2.2.1.
If Assumptions 2.2.1 and 2.2.2 are satisfied, then the economic system has a unique equilibrium. The equilibrium values of the variables are given by the following procedure: $\overline{N}$ by (2.2.13) $\rightarrow$ N_j, $j = 1, 2$, by (2.2.12) $\rightarrow$ K by (2.2.15) $\rightarrow$ K_j by (2.2.10) $\rightarrow$ $r = \alpha \delta_1$ $\rightarrow$ F_j and w_j by (2.2.11) $\rightarrow$ $\bar{r}$ by (2.2.6) $\rightarrow$ k_j by (2.2.7) $\rightarrow$ y_j by (2.1.4) $\rightarrow$ R_j by (2.2.3) $\rightarrow$ l_j, c_j and s_j by (2.2.2) $\rightarrow$ U_j by (2.1.5).

We thus solved the equilibrium problem under Assumptions 2.2.1 and 2.2.2. In order to derive the stability conditions, it is necessary to express Ω_j in (2.1.8) as functions of $k_1(t)$ and $k_2(t)$ at any point of time. This further requires that we have to express all the variables as functions of $k_1(t)$ and $k_2(t)$ at any point of time. Since it is difficult to do this, we omit the issue.

2.3 Changes in Region 1's Working Efficiency

This section examines effects of changes in region 1's working efficiency on the equilibrium structure. Assumptions 2.2.1 and 2.2.2 are assumed to be held. Taking derivatives of (2.2.13) with respect to z_1 yields in which $d\Phi / d\bar{N} < 0$ as given in (2.2.14). By (2.3.1), we have $d\bar{N} / dz_1 > 0$. By (2.2.12), we get

$$\frac{1}{N_1}\frac{dN_1}{dz_1} = \frac{1}{(1+\bar{N})\bar{N}}\frac{d\bar{N}}{dz_1} > 0, \quad \frac{1}{N_2}\frac{dN_2}{dz_1} = -\frac{1}{1+\bar{N}}\frac{d\bar{N}}{dz_1} < 0. \tag{2.3.2}$$

An increase in region 1's working efficiency will make some people migrate from region 2 to region 1.

Taking derivatives of (2.2.6) and (2.2.15) with respect to z_1 yields

$$\frac{1}{\bar{r}}\frac{d\bar{r}}{dz_1} = \frac{1}{K}\frac{dK}{dz_1} = \frac{N_1}{N^*} + \frac{(z_1 - z_2)N_2}{(1+\bar{N})N^*}\frac{d\bar{N}}{dz_1}. \tag{2.3.3}$$

If $z_1 \geq z_2$, as region 1's working efficiency is improved, the total capital of the economy is increased. In the case of $z_1 < z_2$, then K may be either increased or decreased as z_1 is increased. This conclusion is intuitively acceptable since if region 1 has higher level of human capital, then the migration from region 2 to region 1 will increase the total capital. On the other hand, if people in a more efficient region to the other one, then one may expect a reduction in the total capital. By (2.2.10) and (2.2.11), we have

$$\frac{1}{F_1}\frac{dF_1}{dz_1} = \frac{1}{K_1}\frac{dK_1}{dz_1} = \frac{N_1}{N^*} + \frac{N_2}{N}\frac{d\bar{N}}{dz_1} + \frac{1}{(1+z\bar{N})z_1 z_2} > 0,$$
$$\frac{1}{F_2}\frac{dF_2}{dz_1} = \frac{1}{K_2}\frac{dK_2}{dz_1} = -\frac{1}{1+\bar{N}}\frac{d\bar{N}}{dz_1} < 0,$$
$$\frac{1}{w_1}\frac{dw_1}{dz_1} = \frac{N_1}{N^*} + \frac{(N_1 - N_2)(z_1 - z_2)}{(1+\bar{N})\bar{N}N^*}\frac{d\bar{N}}{dz_1} + \frac{1}{(1+z\bar{N})z_1 z_2} > 0$$
$$\frac{dw_2}{dz_1} = \frac{dr}{dz_1} = 0. \tag{2.3.4}$$

Region 1's (region 2's) total output and its total capital stocks are increased (reduced) as region 1's working force becomes more effective. In the case of $(N_1 - N_2)(z_1 - z_2) \geq 0$, region 1's wage rate is increased. In the case of $(N_1 - N_2)(z_1 - z_2) < 0$, region 1's wage rate may be either increased or reduced. Region 2's wage rate w_2 and the rate of interest r are not affected in changes in z_1.

By (2.2.7) and (2.2.2), we get

$$(a-r)\frac{dk_1}{dz_1} = \frac{dw_1}{dz_1} + n\frac{dK}{dz_1},$$
$$(a-r)\frac{dk_2}{dz_1} = \frac{nN_1K}{N^*} + \frac{(z_1 - z_2)nKN_2}{(1+\overline{N})N^*}\frac{d\overline{N}}{dz_1},$$
$$\frac{1}{R_1}\frac{dR_1}{dz_1} = \frac{1}{(1+\overline{N})\overline{N}}\frac{d\overline{N}}{dz_1} + \frac{1}{k_1}\frac{dk_1}{dz_1},$$
$$\frac{1}{R_2}\frac{dR_2}{dz_1} = \frac{1}{k_2}\frac{dk_2}{dz_1} - \frac{1}{1+\overline{N}}\frac{d\overline{N}}{dz_1}. \tag{2.3.5}$$

In the case of $N_1 \geq N_2$ and $z_1 \geq z_2$, region 1's capital stocks per capita, k_1, is increased; otherwise k_1 may be either increased or reduced. In the case of $z_1 \geq z_2$, k_2 is increased; otherwise k_2 may be either increased or reduced. Region 1's land rent, R_1, is increased if its residents' capital stocks per capita is increased; otherwise the sign of dR_1 / dz_1 depends on the comparative force that tends to reduce the reduced living standard and to increase the land rent due to migration. Immigration does not necessarily cause the land rent to increase in our model because a change in labor force's spatial location implies changes in regional output levels and living standard. Similarly, we see that region 2's land rent R_2 may be either increased or reduced.

By $y_j = \delta_1 k_j$, $c_j = \xi k_j / \lambda$ and $l_j = L_j / N_j$, we have

$$\frac{1}{y_j}\frac{dy_j}{dz_1} = \frac{1}{c_j}\frac{dc_j}{dz_1} = \frac{1}{k_j}\frac{dk_j}{dz_1}, \quad \frac{1}{l_1}\frac{dl_1}{dz_1} = -\frac{1}{N_1}\frac{dN_1}{dz_1} < 0,$$
$$\frac{1}{l_2}\frac{dl_2}{dz_1} = -\frac{1}{N_2}\frac{dN_2}{dz_1} > 0. \tag{2.3.6}$$

The effects on region j's income per capita y_j and consumption level per capita c_j are in the same direction as on the capital stocks per capita k_j. Since the territory size of each region is fixed and people migrate from region 2 to region 1, region 1's (2's) lot size per capita is obviously reduced (increased).

2.4 Changes in Region 2's Amenity Level

This section is concerned with the impact of changes in region 2's amenity level A_2 on the economic structure. Taking derivatives of (2.2.13) with respect to A_2 yields

$$-\frac{d\Phi}{d\bar{N}}\frac{d\bar{N}}{dA_2} = -\frac{A\bar{N}^{\eta\theta}}{A_2} < 0, \tag{2.4.1}$$

where $d\Phi / d\bar{N} < 0$. An increase in region 2's amenity level reduces $\bar{N}$. By (2.2.12) we obtain

$$\frac{1}{N_1}\frac{dN_1}{dA_2} = \frac{1}{(1+\bar{N})\bar{N}}\frac{d\bar{N}}{dA_2} < 0, \quad \frac{1}{N_2}\frac{dN_2}{dA_2} = -\frac{1}{1+\bar{N}}\frac{d\bar{N}}{dA_2} > 0. \tag{2.4.2}$$

An improvement in region 2's environment drives some people from region 1 to region 2.

From (2.2.6) and (2.2.15), we get

$$\frac{1}{\bar{r}}\frac{d\bar{r}}{dA_2} = \frac{1}{K}\frac{dK}{dA_2} = \frac{(z_1 - z_2)N_1}{(1+\bar{N})N^*}\frac{d\bar{N}}{dA_2}. \tag{2.4.3}$$

We see that if $z_1 > (<)\ z_2$, then $dK / dA_2 < (>)\ 0$. If region 1's working efficiency is higher (lower) than region 2, then as region 2's amenity level is increased, the total capital of the economy K and per capita's income from land revenue $\bar{r}$ are reduced (increased). By (2.2.10) and (2.2.11), we have

$$\frac{1}{F_1}\frac{dF_1}{dA_2} = \frac{1}{K_1}\frac{dK_1}{dA_2} = \frac{(z_1 - z_2)\bar{N}N_1 + z_2 N}{(1+\bar{N})\bar{N}N^*}\frac{d\bar{N}}{dA_2},$$
$$\frac{1}{F_2}\frac{dF_2}{dA_2} = \frac{1}{K_2}\frac{dK_2}{dA_2} = -\frac{N_1}{N}\frac{d\bar{N}}{dA_2} > 0. \tag{2.4.4}$$

If $z_1 \geq z_2$, then $dK_1 / dA_2 < 0$ and $dF_1 / dA_2 < 0$. If $z_1 < z_2$, then K_1 and F_1 may be either increased or reduced. As region 2 's amenity level is increased, region 2 's total output and capital stocks employed by region 2 are increased.

From (2.2.11), we obtain

$$\frac{1}{w_1}\frac{dw_1}{dA_2} = \frac{(N_1 - N_2)(z_1 - z_2)N_2}{N_1 N^*}\frac{d\overline{N}}{dA_2},$$

$$\frac{1}{w_2}\frac{dw_2}{dA_2} = \frac{(N_1 - N_2)(z_1 - z_2)N_2}{NN^*}\frac{d\overline{N}}{dA_2}, \quad \frac{dr}{dA_2} = 0. \tag{2.4.5}$$

The two regions' wage rates w_j are increased (reduced) in the case of $(N_1 - N_2)(z_1 - z_2) < (>)\ 0$. If $(N_1 - N_2)(z_1 - z_2) = 0$, then w_j, $j = 1, 2$, are not affected. The rate of interest r is not affected by changes in A_2. Taking derivatives of (2.2.7) with respect to A_2 yields

$$(a - r)\frac{dk_j}{dA_2} = \frac{dw_j}{dA_2} + n\frac{dK}{dA_2}. \tag{2.4.6}$$

By (2.4.4) and (2.4.5), we conclude that in the case of $z_1 = z_2$, k_j are not affected. In the case of $N_1 > N_2$ and $z_1 < (>)\ z_2$, k_j are increased; otherwise k_j may be either increased or reduced. By (2.2.3), we have

$$\frac{1}{R_1}\frac{dR_1}{dA_2} = \frac{1}{(1 + \overline{N})\overline{N}}\frac{d\overline{N}}{dA_2} + \frac{1}{k_1}\frac{dk_1}{dA_2},$$

$$\frac{1}{R_2}\frac{dR_2}{dA_2} = \frac{1}{k_2}\frac{dk_2}{dA_2} - \frac{1}{1 + \overline{N}}\frac{d\overline{N}}{dA_2}. \tag{2.4.7}$$

If $dk_1 / dA_2 < 0$, then $dR_1 / dA_2 < 0$; otherwise region 1's land rent R_1 may be either increased or reduced; If $dk_2 / dA_2 > 0$, then $dR_2 / dA_2 > 0$; otherwise region 2 's land rent R_2 may be either increased or reduced.

By $y_j = \delta_1 k_j$, $c_j = \xi k_j / \lambda$ and $l_j = L_j / N_j$, we have

$$\frac{1}{y_j}\frac{dy_j}{dA_2}=\frac{1}{c_j}\frac{dc_j}{dA_2}=\frac{1}{k_j}\frac{dk_j}{dA_2},\ \frac{1}{l_1}\frac{dl_1}{dA_2}=-\frac{1}{N_1}\frac{dN_1}{dA_2}>0,$$
$$\frac{1}{l_2}\frac{dl_2}{dA_2}=-\frac{1}{N_2}\frac{dN_2}{dA_2}<0\ . \tag{2.4.8}$$

The effects on region j 's income per capita y_j and consumption level per capita c_j are in the same direction as on the capital stocks per capita k_j. Since the territory size of each region is fixed and people migrate from region 1 to region 2, region 2 's (1 's) lot size per capita is reduced (increased).

2.5 On Regional Dynamics

We proposed a compact two-region economic model with endogenous capital accumulation. The two regions are different in natural resources (i.e., land in this chapter), amenity (such as climates), working efficiency and life styles. The model describes a dynamic interaction between capital accumulation, regional distribution of capital and labor, regional differences in output levels, land rent and wage rate. We also examined the effects of changes in the amenity level and working efficiency on the equilibrium structure.

It is emphasized that we get explicit conclusions about the effects of changes in some parameters on some special requirements. From our discussion it can be seen that it is difficult to discuss issues related to the existence of economic equilibria and effects of changes in parameters in the cases that the two regions have different elasticities of production and different propensities. We may extend the model in multiple ways.

We neglected issues related to economic production with a refined production structure such as service, manufacturing, consumption goods and tourists sectors, within each region and regional division of production over the entire economy. Conceptually, it is not difficult to extend the discussion to multiple sectors since the multiple sector model of a one-region economy exists in the literature. It is important to introduce infrastructures, population or knowledge dynamics into the system. We may also further disaggregate economic structures and consider geographical structures within each region. It is not difficult to propose some reasonable assumptions to make amenities as endogenous variables represented in the form of

$$\frac{dA_j}{dt}=D_j(F_1,F_2,N_1,N_2,A_1,A_2),\quad j=1,\,2\,.$$

Appendix

A.2.1 Equilibria in the Case that the Propensities to Consume and to Hold Wealth Are Different

The appendix is concerned with conditions for the existence of equilibria of the dynamic system when the two regions have different propensities to hold wealth. For convenience, we require Assumption 2.2.2 to be satisfied.

By (2.1.8), at equilibrium we have

$$\lambda_j \Omega_j = k_j, \quad j = 1, 2. \tag{2.A.1.1}$$

Substituting (2.A.1.1) into (2.1.7) yields

$$l_j = \frac{\eta_j k_j}{\lambda_j R_j}, \; c_j = \frac{\xi_j k_j}{\lambda_j}, \quad s_j = \delta_k k_j. \tag{2.A.1.2}$$

By $l_j = \eta_j k_j / \lambda_j R_j$ in (2.A.1.2) and $l_j N_j = L_j$ in (2.1.11)

$$R_j = \frac{\eta_j k_j N_j}{\lambda_j L_j}. \tag{2.A.1.3}$$

Substituting c_j and s_j in (2.A.1.2) and $l_j N_j = L_j$ in (2.1.11) into (2.1.5) and then using (2.1.9), we get

$$\frac{k_1^{\xi_1 + \lambda_1}}{k_2^{\xi_2 + \lambda_2}} = \frac{A N_1^{\eta_1}}{N_2^{\eta_2}} \tag{2.A.1.4}$$

in which

$$A \equiv \frac{A_2 L_2^{\eta_2} (\xi_2 / \lambda_2)^{\xi_2} (\lambda_1 / \xi_1)^{\xi_1}}{A_1 L_1^{\eta_1}}.$$

By (2.A.1.2) and (2.1.12), we get

$$\delta_1 k_1 N_1 + \delta_2 k_2 N_2 = F_1 + F_2 \tag{2.A.1.5}$$

where $\delta_j \equiv \xi_j / \lambda_j + s_j$.

By (2.1.3) and (2.A.1.3), we get

$$\bar{r} = n_1 k_1 N_1 + n_2 k_2 N_2 \tag{2.A.1.6}$$

where $n_j \equiv \eta_j / \lambda_j N$, $j = 1, 2$. By (2.A.1.1), (2.1.4) and the definitions of W_j, we get

$$w_j + r k_j = a_j k_j - n_1 k_1 N_1 - n_2 k_2 N_2, \quad j = 1, 2 \tag{2.A.1.7}$$

where $a_j \equiv 1/\rho_j \lambda_j - \delta > 0$. By (2.1.1), $r = \alpha_1 F_1 / K_1 = \alpha_2 F_2 / K_2$ and $K_1 + K_2 = K$, we have

$$K_1 = \frac{z\bar{N}K}{1 + z\bar{N}}, \quad K_2 = \frac{K}{1 + z\bar{N}}. \tag{2.A.1.8}$$

where $\bar{N} \equiv N_1 / K_2$ and $z \equiv z_1 / z_2$. By (2.A.1.8), we obtain $z_1 N_1 / K_1 = z_2 N_2 / K_2$. From this equation and $w_j = \beta F_j / N_j = \beta z_j (K_j / z_j N_j)^{\alpha}$, we get

$$\frac{w_1}{w_2} = z, \quad \frac{F_1}{F_2} = \frac{zN_1}{N_2}. \tag{2.A.1.9}$$

By (2.1.2), we get

$$w_j + r k_j = \frac{\beta + \alpha k_j N_j / K_j}{N_j} F_j, \quad j = 1, 2 .$$

Substituting (2.A.1.8) into these two equations yields

$$w_1 + r k_1 = \frac{\beta + \alpha\Lambda(1 + z\bar{N})}{(1 + \Lambda\bar{N}) z N_1} F_1, \quad w_2 + r k_2 = \frac{\beta + \alpha(1 + z\bar{N})}{1 + \Lambda\bar{N}} F_2 \tag{2.A.1.10}$$

in which (2.A.1.8) is used and $\Lambda \equiv k_1 / k_2$. Dividing the first equation in (2.A.1.7) by the second one, we get

$$\frac{\beta z + \alpha\Lambda + z\Lambda\overline{N}}{1 + \alpha z\overline{N} + \beta\Lambda\overline{N}} = \frac{a_1\Lambda - n_1N_1\Lambda - n_2N_2}{a_2 - n_1N_1\Lambda - n_2N_2} \tag{2.A.1.11}$$

in which (2.A.1.11) and (2.A.1.10) are used.

As $k_j > 0$, by (2.A.1.7), we see that it is necessary to require

$$a_1\Lambda > n_1N_1\Lambda + n_2N_2, \quad a_2 > n_1N_1\Lambda + n_2N_2$$

i.e.

$$\frac{a_2 - n_2N_2}{n_1N_1} > \Lambda > \frac{n_2N_2}{a_1 - n_1N_1} \tag{2.A.1.12}$$

where we use $a_j / n_j > N_j$ when $N_j < N,\ j = 1, 2$.

The equation, (2.A.1.11) is rewritten as

$$\Lambda^2 - 2f_1\Lambda - f_2 = 0 \tag{2.A.1.13}$$

in which

$$f_1(N_1, N_2) \equiv$$
$$\frac{a_0(\alpha + z\overline{N}) - (a_1 - n_1N_1)(1 + \alpha z\overline{N}) + \beta n_2N_2\overline{N} - \beta z n_1N_1}{2\left[(\alpha + z\overline{N})n_1N_1 + (a_1 - n_1N_1)\beta\overline{N}\right]},$$
$$f_2(N_1, N_2) \equiv \frac{a_0\beta z + n_2N_2(1 + \alpha z\overline{N})}{(\alpha + z\overline{N})n_1N_1 + (a_1 - n_1N_1)\beta\overline{N}},$$
$$a_0 \equiv a_2 - n_2N_2.$$

As $f_2 > 0$, (2.A.1.13) has a unique positive solution as follows

$$\Lambda = f_1 + \left(f_1^2 + f_2\right)^{1/2}. \tag{2.A.1.14}$$

We thus solve Λ as a unique function of N_1 and N_2. We now show that we can represent k_1 and k_2 as unique functions of N_1 and N_2.

By (2.A.1.5) and (2.A.1.9), we have

$$\left(\delta_1 \Lambda \overline{N} + \delta_2\right) k_2 = (1 + z\overline{N}) \frac{F_2}{N_2}.$$

By (2.1.1), $K_2 = K/(1 + z\overline{N})$ and (2.1.10), we have

$$\frac{F_2}{N_2} = z_2^{\beta} k_2^{\alpha} \left(\frac{1 + \Lambda \overline{N}}{1 + z\overline{N}} \right)^{\alpha}.$$

By these two equations and $\Lambda = k_1 / k_2$, we solve

$$k_1 = k_2 \Lambda, \quad k_2 = \frac{z_2 \left(1 + z\overline{N}\right)\left(1 + \Lambda \overline{N}\right)^{\alpha/\beta}}{\left(\delta_2 + \delta_1 \Lambda \overline{N}\right)^{1/\beta}}. \tag{2.A.1.15}$$

We have thus solved k_1 and k_2 as unique functions of N_1 and N_2. We now solve N_1 and N_2.

By $N_1 + N_2 = N$ and $\overline{N} = N_1 / N_2$, we have

$$N_1 = \frac{N\overline{N}}{1 + \overline{N}}, \quad N_2 = \frac{N}{1 + \overline{N}}. \tag{2.A.1.16}$$

We see that we may treat N_1 and N_2 as functions of $\overline{N}$. Substituting (2.A.1.15) and (2.A.1.16) into

$$\Phi(\overline{N}) \equiv \Lambda(\overline{N})^{\xi_1 + \lambda_1} k_2(\overline{N})^{\xi_0} - A N^{\eta_0} \overline{N}^{\eta_1} (1 + \overline{N})^{-\eta_0} = 0 \tag{2.A.1.17}$$

where

$$\xi_0 \equiv \xi_1 + \lambda_1 - \xi_2 - \lambda_2, \quad \eta_0 \equiv \eta_1 - \eta_2.$$

In (2.A.1.17), Λ and k_2 are unique functions of $\overline{N}$. Hence, (2.A.1.17) includes a single variable $\overline{N}$. Summarizing the above discussion, we obtain the following lemma.

Lemma 2.A.1.1.

Let Assumption 2.2.2 be satisfied. If (2.A.1.17) has positive solutions $\overline{N}$ such that (2.A.1.12) are satisfied, the dynamic system has equilibria. The number of equilibria is equal to the number of the positive solutions satisfying (2.A.1.12). The equilibrium values of the variables are given by the following procedure: $\overline{N}$ by (2.A.1.17) $\rightarrow$ N_j, $j = 1, 2$, by (2.A.1.16) $\rightarrow$ L by (2.A.1.14) $\rightarrow$ k_j by (2.A.1.15) $\rightarrow$ $K = k_1 N_1 + k_2 N_2$ $\rightarrow$ K_j by (2.A.1.8) $\rightarrow$ F_j by (2.1.1) $\rightarrow$ r and w_j by (2.1.2) $\rightarrow$ $\bar{r}$ by (2.A.1.6) $\rightarrow$ y_j by (2.1.4) $\rightarrow$ R_j by (2.A.1.3) $\rightarrow$ l_j, c_j and s_j by (2.A.1.2) $\rightarrow$ U_j by (2.1.5).

3 Regional Growth with Endogenous Time Distribution

Various models were proposed to explain why regions trade and how interregional trade patterns are determined. Different economists emphasize various factors, such as differences in resources, capital accumulation and technology, in explaining trade issues. In Chapter 2, we studied regional growth and interregional trade patterns. The concern of this chapter is still to extend the model developed in Chapter 2. The main difference of this chapter from Chapter 2 is related to behavior of households of different regions. We consider endogenous time in our chapter.

Over the years there have been a number of attempts to modify neoclassical consumer theory to deal with economic issues about endogenous labor supply, family structure, working hours and the valuation of traveling time. There are studies on relationships between home production and non-home production and time distribution (e.g., Becker, 1965, Lancaster, 1966, 1971). This chapter makes an initial attempt to introduce endogenous choice between working and leisure time to the neoclassical dynamic growth theory with interregional trade. This chapter proposes a two-region growth model of endogenous capital accumulation and time under perfect competition and free movement of the population with given regional differences in land, amenity and life styles. The chapter is organized as follows. Section 3.1 defines the basic model of economic growth with endogenous time and interregional trade. Section 3.2 provides the conditions for the existence of equilibrium. Sections 3.3 and 3.4 examine the impact of changes in the work efficiency and the regional amenity levels on the regional economic structure. Section 3.5 concludes the chapter. This chapter is based on Zhang (1995a, 1998a).

3.1 Growth with Regional Time Values

The basic structure of the model in this chapter is similar to the model proposed in Chapter 2. We consider an economic system which consists of two regions, indexed by 1 and 2, respectively. Each region has a single production sector. Production is carried out by combining capital and labor. It is assumed that the two regions' product is qualitatively homogenous and is either consumed or invested. We assume a homogenous population. A person is free to choose where he works and where he lives. We assume that any person chooses the same region where he works and lives.

Each region has fixed land. It is assumed that land quality, climates and environment are homogenous within each region, but climates and environment may be different between the two regions. The land is used only for housing. We select commodity to serve as numeraire, with all the other prices being measured relative to its price. We neglect transportation cost of commodities between and within regions.

We introduce

N — the given population of the economy;

L_j — the given territory size of region j, $j = 1, 2$;

$K(t)$ — the total capital stocks of the economy at time t;

$F_j(t)$ — the output levels of region j 's industrial sector at time t;

$K_j(t)$ and $N_j(t)$ — the levels of capital stocks and labor force employed by region j 's production sector;

$c_j(t)$ and $s_j(t)$ — per capita's consumption level of commodity and savings made by per capita in region j;

$l_j(t)$ — the lot size per capita in region j;

$T_j(t)$ and $T_j^h(t)$ — the working time and leisure time of a worker in region j;

$y_j(t)$ — the net income per capita in region j;

$r(t)$ — the rate of interest; and

$w_j(t)$ and $R_j(t)$ — region j 's wage rate per unity of working time and land rent, respectively.

We now describe the model.

The production sectors

We specify the production functions of the two regions as follows

$$F_j(t) = K_j^{\alpha}\left(z_j T_j N_j\right)^{\beta}, \quad \alpha, \beta > 0, \quad \alpha + \beta = 1, \quad j = 1, 2 \qquad (3.1.1)$$

in which z_j are working efficiency index of region j 's labor force and $z_j T_j N_j$ is region j 's total efficient labor force. For simplicity, we assume z_j to be constant. In this chapter we assume $z_1 > z_2$. That is, region 1 's working force is more effective than region 2 's labor force.

The marginal conditions are given by

$$r = \frac{\alpha F_j}{K_j}, \quad w_j = \frac{\beta F_j}{T_j N_j}, \quad j = 1, 2. \tag{3.1.2}$$

The rate of interest is identical in the whole economy and the wage rates, w_j, may be different between the two regions.

Behavior of households

Each worker may get income from land ownership, capital ownership and wages. In order to define net incomes, it is necessary to determine land ownership structure. It can be seen that land properties may be distributed in multiple ways under various institutions. To simplify the discussion, we specify land revenue as follows. We assume that each worker owns L_1 / N amount of land in region 1 and L_2 / N in region 2 and it is impossible to sell land but it is free to rent one's own land to others. Under this assumption, the land revenue, R_0, per worker is given as follows

$$R_0(t) = \frac{L_1 R_1(t) + L_2 R_2(t)}{N}. \tag{3.1.3}$$

This assumption implies that each worker receives the land revenue, R_0, irrespective of whether the worker's dwelling site is located in region 1 or region 2.

If we denote $k_j(t)$ capital stocks owned by per capita in region j, the interest payment per capita is given by $r(t)k_j(t)$. Under the specified land ownership, the net income per capita, $y_j(t)$, in region j consists of the wage income, $T_j(t)w_j(t)$, land revenue, R_0, and interest payment $r(t)k_j(t)$, i.e.

$$y_j(t) = T_j(t)w_j(t) + r(t)k_j(t) + R_0(t). \tag{3.1.4}$$

It is assumed that there is an aggregated utility function for each region. It is assumed that a typical person's utility level, $U_j(t)$, in region j is dependent on the leisure time, $T_j^h(t)$, lot size, $l_j(t)$, consumption level, $c_j(t)$, of community, and the level of the person's wealth, $k_j(t) + s_j(t) - \delta_k k_j(t)$, where δ_k is the fixed depreciation rate of capital. The utility functions are specified as follows

$$U_j(t) = A_j T_j^h(t)^{\sigma} l_j(t)^{\eta} c_j(t)^{\xi} \{k_j(t) + s_j(t)^{\eta} - \delta_k k_j(t)\}^{\lambda},$$
$$\sigma, \eta, \xi, \lambda > 0, \quad \sigma + \eta + \xi + \lambda = 1 \tag{3.1.5}$$

in which σ, η, ξ, and λ are respectively the household's propensity to use leisure time, to utilize lot size, to consume the commodity, and to hold wealth. In (3.1.5), A_j is called region j's amenity level. In this chapter, the amenity levels are considered as locational parameters or slow changing variables, such as infrastructures, regional cultures and climates, which may affect attractiveness of a location. Here, we neglect possible dynamics of amenity.

Let T_0 denote the total available time. The time constraint requires that the amounts of time allocated to each specific use add up to the time available

$$T_j + T_j^h = T_0, \quad j = 1, 2. \tag{3.1.6}$$

The budget constraints are given by

$$l_j R_j + c_j + s_j = y_j, \quad j = 1, 2. \tag{3.1.7}$$

By (3.1.4) and (3.1.6), we rewrite (3.1.7) as follows

$$l_j R_j + c_j + s_j + T_j w_j = R_0 + T_j^h w_j + r k_j, \quad j = 1, 2. \tag{3.1.8}$$

A typical person maximizes the person's utility subject to (3.1.8). The optimal problems have the following solutions

$$T_j^h w_j = \sigma \Omega_j, \quad l_j R_j = \eta \Omega_j, \quad c_j = \xi \Omega_j, \quad s_j = \lambda \Omega_j - \delta k_j, \quad j = 1, 2 \tag{3.1.9}$$

where

$$\Omega_j \equiv R_0 + T_0 w_j + r k_j + \delta k_j, \quad \delta \equiv 1 - \delta_k.$$

According to the definitions of k_j and s_j, the capital accumulation of a typical person in region j is given by

$$\frac{dk_j}{dt} = s_j - \delta_k k_j.$$

Substituting s_j in (3.1.9) into these two equations yields

$$\frac{dk_j}{dt} = \lambda \Omega_j - k_j, \quad j = 1, 2. \tag{3.1.10}$$

As households are freely mobile between the two regions, they should have the same level of the utility, irrespective of where they live. That is

$$U_1(t) = U_2(t). \tag{3.1.11}$$

Balance of demand and supply

By the definitions of K, k_j and N_j

$$K = k_1 N_1 + k_2 N_2. \tag{3.1.12}$$

The assumption that labor force, capital stocks and land are fully employed is represented by

$$N_1 + N_2 = N, \quad K_1 + K_2 = K, \quad l_j N_j = L_j, \quad j = 1, 2. \tag{3.1.13}$$

The balance of demand for and supply of industrial commodity is given by

$$(c_1 + s_1) N_1 + (c_2 + s_2) N_2 = F_1 + F_2. \tag{3.1.14}$$

We have thus built the model. The system has 29 variables, N_j, K_j, F_j, c_j, s_j, k_j, l_j, T_j, T_j^h, y_j, U_j, w_j, R_j ($j = 1, 2$), K, r and R_0. It contains the same number of independent equations. We now examine conditions for the existence of equilibria of the dynamic system.

3.2 The Equilibrium Structure with Identical Propensities

This section is concerned with the conditions for existence of economic equilibria. By (3.1.10), at equilibrium we have

$$\lambda \Omega_j = k_j, \quad j = 1, 2. \tag{3.2.1}$$

Substituting (3.2.1) into (3.1.9) yields

$$T_j^h w_j = \frac{\sigma k_j}{\lambda}, \quad l_j R_j = \frac{\eta k_j}{\lambda}, \quad c_j = \frac{\xi k_j}{\lambda}, \quad s_j = \delta_k k_j, \quad j = 1, 2. \tag{3.2.2}$$

Substituting (3.2.2) into (3.1.14) yields

$$\delta_1 K = F_1 + F_2 \tag{3.2.3}$$

where we use (3.1.12) and

$$\delta_1 \equiv \frac{\xi}{\lambda} + \delta_k .$$

By $r = \alpha F_1 / K_1 = \alpha F_2 / K_2$ in (3.1.2) and (3.2.3), we have

$$r = \alpha \delta_1 \tag{3.2.4}$$

where we use $K_1 + K_2 = K$. By (3.1.1), $r = \alpha F_1 / K_1 = \alpha F_2 / K_2$ in (3.1.2) and (3.2.4), we get

$$z_j T_j N_j = \delta_1^{1/\beta} K_j, \quad j = 1, 2. \tag{3.2.5}$$

By (3.2.5) and (3.1.2)

$$w_j = \frac{\beta z_j}{\delta_1^{\alpha/\beta}}, \quad j = 1, 2. \tag{3.2.6}$$

Substituting $l_j = L_j / N_j$ and (3.2.2) into (3.1.5) and then using (3.1.11), we obtain

$$\frac{k_1}{k_2} = A \Lambda^{\eta \upsilon} \tag{3.2.7}$$

where we use $w_1 / w_2 = z$ and

$$\Lambda \equiv \frac{N_1}{N_2}, \quad z \equiv \frac{z_1}{z_2}, \quad \upsilon \equiv \frac{1}{1 - \eta}, \quad A \equiv z^{\sigma \upsilon} \left(\frac{A_2}{A_1} \right)^{\upsilon} \left(\frac{L_2}{L_1} \right)^{\upsilon \eta}.$$

By the definition of Λ and $N_1 + N_2 = N$, we have

$$N_1 = \frac{\Lambda N}{1 + \Lambda}, \quad N_2 = \frac{N}{1 + \Lambda}. \tag{3.2.8}$$

By $l_j R_j = \eta k_j / \lambda$ in (3.2.2), $l_j = L_j / N_j$ and (3.1.3), we get

$$R_0 = \frac{\eta K}{\lambda N} \tag{3.2.9}$$

where we use (3.1.12).

By the definitions of Ω_j and (3.2.1), we get

$$R_0 + T_0 w_j = \delta_2 k_j, \quad j = 1, 2 \tag{3.2.10}$$

where

$$\delta_2 \equiv \frac{1}{\lambda} - \delta - r = \frac{\sigma + \eta + \beta\xi}{\lambda} + \beta\delta_k > 0.$$

By (3.2.10), we have

$$k_1 - k_2 = w_0 \equiv (z - 1)\frac{T_0 w_2}{\delta_2} > 0. \tag{3.2.11}$$

Region 1's wage rate is higher than region 2's wage rate, i.e., $w_1 > w_2$ and the level of capital stocks owned by per capita in region 1 is higher than that in region 2, i.e., $k_1 > k_2$. By (3.2.7) and (3.2.11), we solve

$$k_1 = \frac{w_0 A \Lambda^{\eta\upsilon}}{A\Lambda^{\eta\upsilon} - 1}, \quad k_2 = \frac{w_0}{A\Lambda^{\eta\upsilon} - 1}. \tag{3.2.12}$$

It is necessary to require

$$A\Lambda^{\eta\upsilon} > 1.$$

Substituting (3.2.9) and k_2 in (3.2.12) into $R_0 + T_0 w_2 = \delta_2 k_2$ in (3.2.10), we have

$$K = \frac{\lambda N T_0 w_2}{\eta\left(A\Lambda^{\eta\upsilon} - 1\right)}\left(z - A\Lambda^{\eta\upsilon}\right). \tag{3.2.13}$$

Since $A\Lambda^{\eta\upsilon} > 1$, for $K > 0$ it is necessary to require: $z > A\Lambda^{\eta\upsilon}$. Since $k_j > 0$ are guaranteed by $A\Lambda^{\eta\upsilon} > 1$, by $T_j^h w_j = \sigma k_j / \lambda$ in (3.2.2), we see that $T_0 > T_j^h > 0$ are satisfied if $T_0 > \sigma k_j / \lambda w_j$, $j = 1, 2$. By (3.2.12), we see that these two inequalities are satisfied if

$$A\Lambda^{\eta\upsilon} > \frac{\lambda\delta_2 z}{\lambda\delta_2 z - \sigma(z - 1)}, \quad A\Lambda^{\eta\upsilon} > \frac{\sigma}{\lambda\delta_2}(z - 1) + 1.$$

Since

$$z > \frac{\sigma}{\lambda\delta_2}(z - 1) + 1 > \frac{\lambda\delta_2 z}{\lambda\delta_2 z - \sigma(z - 1)} > 1$$

we should require Λ to satisfy

$$z > A\Lambda^{\eta\upsilon} > \frac{\sigma}{\lambda\delta_2}(z - 1) + 1. \tag{3.2.14}$$

Under (3.2.14), we have $K > 0$, $k_j > 0$ and $T_0 > T_j^h > 0$.

Substituting (3.2.8) and (3.2.12) into (3.1.12) and then using (3.2.13), we get

$$\Phi(\Lambda) \equiv \left(A\Lambda^{\eta} + 1\right)(z - 1) - \left(z - A\Lambda^{\eta\upsilon}\right)(1 + \Lambda)\frac{\lambda\delta_2}{\eta}. \tag{3.2.15}$$

It is shown that the function, $\Phi(\Lambda)$, has the following properties

$$\Phi = \frac{\beta\xi + \beta\lambda\delta_k - (z-1)\sigma\eta/\lambda\delta_2}{\eta}(1+\Lambda)(z-1)$$

$$-\frac{\sigma}{\lambda\delta_2}(z-1)^2, \quad at \ A\Lambda^{\eta\upsilon} = \frac{\sigma}{\lambda\delta_2}(z-1)+1,$$

$$\Phi = (1+z\Lambda)(z-1) \quad at \ A\Lambda^{\eta\upsilon} = z,$$

$$\Phi^* \equiv \frac{d\Phi}{d\Lambda} = \upsilon A\Lambda^{\eta\upsilon}(z-1) + \upsilon\lambda\delta_2 A\Lambda^{\eta\upsilon-1}(1+\Lambda) - \frac{z\lambda\delta_2}{\eta}$$

$$+\frac{\lambda\delta_2 A\Lambda^{\eta\upsilon}}{\eta} > \left\{\frac{\sigma\upsilon}{\lambda\delta_2}(z-1) + \upsilon + \frac{\sigma\upsilon}{\eta} - \frac{\lambda\delta_2}{\eta}\right\}(z-1)$$

$$+\upsilon\lambda\delta_2, \quad \text{for } z > A\Lambda^{\eta\upsilon} > \frac{\sigma}{\lambda\delta_2}(z-1)+1. \tag{3.2.16}$$

Since $\Phi > 0$ at $A\Lambda^{\eta\upsilon} = z$, we see that the equation, $\Phi(\Lambda) = 0$, has at least one solution satisfying (3.2.14) if $\Phi < 0$ at $A\Lambda^{\eta\upsilon} = (z-1)\sigma/\lambda\delta_2 + 1$. By the first equation in (3.2.16), we see that $\Phi < 0$ at $A\Lambda^{\eta\upsilon} = (z-1)\sigma/\lambda\delta_2 + 1$ is held if $\xi + \lambda\delta_k > (z-1)\sigma\eta/\lambda\beta\delta_2$. As $\sigma\eta/\lambda\delta_2 < \min\{\sigma, \eta\}$, we see that if $z-1$ is not large, the requirement is satisfied. If $\Phi^* > 0$ under (3.2.14), the equation has a unique solution in the interval. By (3.2.16), we see that it is acceptable to require $\Phi^* > 0$. For instance, if $\eta + (z-1)(\sigma+\eta)/\lambda\delta_2 > (z-1)/\upsilon$, then $\Phi^* > 0$ under (3.2.14) is held.

Proposition 3.2.1.
We assume

$$\xi + \lambda\delta_k > (z-1)\frac{\sigma\eta}{\lambda\beta\delta_2}, \quad \eta + \frac{\sigma+\eta}{\lambda\delta_2}(z-1) > \frac{z-1}{\upsilon}.$$

The dynamic system has a unique equilibrium. The unique equilibrium values of the variables are given by the following procedure: r by (3.2.4) $\rightarrow$ w_j, $j = 1, 2$, by (3.2.6) $\rightarrow$ R_0 by (3.2.15) $\rightarrow$ N_j by (3.2.8) $\rightarrow$ k_j by (3.2.12) $\rightarrow$ K by (3.2.13) $\rightarrow$ T_j^h by (3.2.9) $\rightarrow$ T_j, $0 < T_j^h < T_0$, by (3.2.2) $\rightarrow$ $T_j = T_0 - T_j^h$ $\rightarrow$ K_j by (3.2.5) $\rightarrow$ F_j by (3.1.1) $\rightarrow$ $l_j = L_j/N_j \rightarrow$ R_j, c_j and s_j by (3.2.2) $\rightarrow$ y_j by (3.1.4) $\rightarrow$ U_j by (3.1.5).

The assumptions in Proposition 3.2.1 are satisfied if the difference between the two regions' levels of working efficiency is appropriately small. It should be remarked that the dynamic system might have a unique equilibrium even if the requirements in Proposition 3.2.1 are not satisfied. In the remainder of this chapter, we examine effects of changes in some parameters. We accept the assumptions in Proposition 3.2.1 in the remainder of the chapter.

3.3 The Working Efficiency and the Equilibrium Structure

This section is concerned with the effects of changes in region 1's working efficiency, z_1, on the economic structure. Taking derivatives of (3.2.15) with respect to z_1 yields

$$z_1 \Phi^* \frac{d\Lambda}{dz_1} = (1 - z)\sigma\upsilon A \Lambda^{\upsilon} - zA\Lambda^{\upsilon} + \frac{\sigma + \beta\xi + \beta\lambda\delta_k}{\eta} z + \frac{z\lambda\delta_2\Lambda}{\eta}$$

$$- \frac{\sigma\lambda\delta_2\upsilon A\Lambda^{\eta\upsilon}}{\eta} - \frac{\sigma\lambda\delta_2\upsilon A\Lambda^{\upsilon}}{\eta} > \frac{\sigma + \beta\xi + \beta\lambda\delta_k - \sigma\lambda\delta_2\upsilon}{\eta} z$$

$$+ \left\{(1 - \sigma\upsilon)\frac{z\lambda\delta_2}{\eta} + \sigma\upsilon - \sigma\upsilon z^2 - z^2\right\}\Lambda \qquad (3.3.1)$$

in which we use (3.2.14) to get the right-hand side of (3.3.1) and $\Phi^* > 0$ under Proposition 3.2.1. We have

$$\frac{\sigma + \beta\xi + \beta\lambda\delta_k - \sigma\lambda\delta_2\upsilon}{\upsilon} = (1 - \sigma - 2\eta)\sigma + (1 - \sigma - \eta)\beta\lambda\delta_k$$

$$+ (1 - \sigma - \eta)\beta\xi.$$

As $\sigma + \eta + \xi + \lambda = 1$, the above term is positive if $\xi + \lambda > \eta$. By (3.3.1), we see that if the term

$$(1 - \sigma\upsilon)\frac{z\lambda\delta_2}{\eta} + \sigma\upsilon - \sigma\upsilon z^2 - z^2 > 0$$

then $d\Lambda / dz_1 > 0$. As

$$(1-\sigma\upsilon)\frac{z\lambda\delta_2}{\eta}+\sigma\upsilon-\sigma\upsilon z^2-z^2=\left\{(\xi+\lambda)\frac{\upsilon\lambda\delta_2}{\eta}-z\right\}z$$
$$-\sigma\upsilon(1+z)(z-1)$$

we see that the term is positive if $z-1$ and η are sufficiently small. Otherwise, it is difficult to judge the sign of the term. Summarizing the above discussion, we conclude that if $z-1$ and η are sufficiently small, it is reasonable to have $d\Lambda / dz_1 > 0$. In the remainder of this section, we assume $d\Lambda / dz_1 > 0$. We have

$$\frac{1}{N_1}\frac{dN_1}{dz_1}=\frac{1}{(1+\Lambda)\Lambda}\frac{d\Lambda}{dz_1}>0,\quad \frac{1}{N_2}\frac{dN_2}{dz_1}=-\frac{1}{1+\Lambda}\frac{d\Lambda}{dz_1}<0. \tag{3.3.2}$$

If the regional working efficiency difference, $z-1$ and the propensity η to consume housing are appropriately small, some of region 2's population will migrate to region 1 as region 1's working efficiency is improved.

By (3.1.7) and (3.2.2), $y_j=(\eta/\lambda+\eta/\lambda+\delta_k)k_j$. By (3.2.12) and these two equations, we have

$$\frac{1}{y_1}\frac{dy_1}{dz_1}=\frac{1}{k_1}\frac{dk_1}{dz_1}=\frac{1}{(z-1)z_2}-\frac{\sigma\upsilon}{(A\Lambda^{\eta\upsilon}-1)z_1}-\frac{\eta\upsilon}{(A\Lambda^{\eta\upsilon}-1)\Lambda}\frac{d\Lambda}{dz_1},$$
$$\frac{1}{y_2}\frac{dy_2}{dz_1}=\frac{1}{k_2}\frac{dk_2}{dz_1}=\frac{1}{(z-1)z_2}-\frac{\sigma\upsilon A\Lambda^{\eta\upsilon}}{(A\Lambda^{\eta\upsilon}-1)z_1}-\frac{\eta\upsilon A\Lambda^{\eta\upsilon-1}}{(A\Lambda^{\eta\upsilon}-1)\Lambda}\frac{d\Lambda}{dz_1}. \tag{3.3.3}$$

It is difficult to explicitly judge the signs of dk_j / dz_1. By (3.2.4) and (3.2.6), we have

$$\frac{dr}{dz_1}=\frac{dw_2}{dz_1}=0,\quad \frac{dw_1}{dz_1}=\frac{w_1}{z_1}>0. \tag{3.3.4}$$

By (3.2.9) and (3.2.10), we get

$$\frac{1}{R_0}\frac{dR_0}{dz_1}=\frac{1}{K}\frac{dK}{dz_1}=\frac{\lambda\delta_2 N}{\eta K}\frac{dk_2}{dz_1}. \tag{3.3.5}$$

By $l_j = L_j / N_j$

$$\frac{1}{l_1}\frac{dl_1}{dz_1} = -\frac{1}{N_1}\frac{dN_1}{dz_1} < 0, \quad \frac{1}{l_2}\frac{dl_2}{dz_1} = -\frac{1}{N_2}\frac{dN_2}{dz_1} > 0. \tag{3.3.6}$$

By (3.2.2) and $T_j + T_j^h = T_0$, we get

$$\frac{1}{c_j}\frac{dc_j}{dz_1} = \frac{1}{k_j}\frac{dk_j}{dz_1}, \quad \frac{dT_j}{dz_1} = -\frac{1}{k_j}\frac{dT_j^h}{dz_1}, \quad j = 1, 2,$$
$$\frac{1}{T_1^h}\frac{dT_1^h}{dz_1} = \frac{1}{k_1}\frac{dk_1}{dz_1} - \frac{1}{z_1}, \quad \frac{1}{T_2^h}\frac{dT_2^h}{dz_1} = \frac{1}{k_2}\frac{dk_2}{dz_1}. \tag{3.3.7}$$

By (3.1.1), (3.2.5), $R_j = \eta K_j / \lambda L_j$ and $K_j = k_j N_j$, we have

$$\frac{1}{R_j}\frac{dR_j}{dz_1} = \frac{1}{F_j}\frac{dF_j}{dz_1} = \frac{1}{K_j}\frac{dK_j}{dz_1} = \frac{1}{k_j}\frac{dk_j}{dz_1} + \frac{1}{N_j}\frac{dN_j}{dz_1}, \quad j = 1, 2. \tag{3.3.8}$$

3.4 The Regional Amenity and the Equilibrium Structure

This section is concerned with effects of changes in amenity levels of the two regions. By (3.2.15) and (3.2.8), we have

$$\Phi^* \frac{d\Lambda}{dA} = (1 - z)\Lambda^{\upsilon} - \frac{\lambda \delta_2 \Lambda^{\eta\upsilon}}{\eta}(1 + \Lambda) < 0,$$
$$\frac{1}{N_1}\frac{dN_1}{dA} = \frac{1}{(1 + \Lambda)\Lambda}\frac{d\Lambda}{dA} < 0, \quad \frac{1}{N_2}\frac{dN_2}{dA} = -\frac{1}{1 + \Lambda}\frac{d\Lambda}{dA} > 0. \tag{3.4.1}$$

As

$$A \equiv z^{\sigma\upsilon}\left(\frac{A_2}{A_1}\right)^{\upsilon}\left(\frac{L_2}{L_1}\right)^{\upsilon\eta}$$

a change in A may be caused by changes in working efficiency, amenity levels and land. In this section, we require that a change in A is caused by changes in A_2 or A_1. Hence, an increase in A may be interpreted as an improvement in region 2's amenity level or a decrease in region 1's amenity level. As region 2's amenity is improved, some of region 1's population will migrate to region 2.

By $y_j = (\xi / \lambda + \eta / \lambda + \delta_k)k_j$ and (3.2.12), we have

$$\frac{1}{y_1}\frac{dy_1}{dA} = \frac{1}{k_1}\frac{dk_1}{dA} = -\frac{1}{(A\Lambda^{\eta\upsilon} - 1)A} - \frac{\eta\upsilon A\Lambda^{\eta\upsilon-1}}{A\Lambda^{\eta\upsilon} - 1}\frac{d\Lambda}{dA},$$
$$\frac{1}{y_2}\frac{dy_2}{dA} = \frac{1}{k_2}\frac{dk_2}{dA} = -\frac{\Lambda^{\eta\upsilon}}{A\Lambda^{\eta\upsilon} - 1} - \frac{\eta\upsilon A\Lambda^{\eta\upsilon-1}}{A\Lambda^{\eta\upsilon} - 1}\frac{d\Lambda}{dA}. \tag{3.4.2}$$

By (3.2.4) and (3.2.6), we have

$$\frac{dr}{dA} = \frac{dw_j}{dA} = 0, \quad j = 1, 2. \tag{3.4.3}$$

The rate of interest and the two regions' wage rates are not affected. By (3.2.9) and (3.2.10), we get

$$\frac{1}{R_0}\frac{dR_0}{dA} = \frac{1}{K}\frac{dK}{dA} = \frac{\lambda\delta_2 N}{\eta K}\frac{dk_2}{dA}. \tag{3.4.4}$$

By $l_j = L_j / N_j$, we have

$$\frac{1}{l_1}\frac{dl_1}{dA} = -\frac{1}{N_1}\frac{dN_1}{dA} > 0, \quad \frac{1}{l_2}\frac{dl_2}{dA} = -\frac{1}{N_2}\frac{dN_2}{dA} < 0. \tag{3.4.5}$$

By (3.2.2) and $T_j + T_j^h = T_0$, we have

$$\frac{1}{c_j}\frac{dc_j}{dA} = \frac{1}{k_j}\frac{dk_j}{dA}, \quad \frac{dT_j}{dA} = -\frac{1}{k_j}\frac{dT_j^h}{dA},$$
$$\frac{1}{T_j^h}\frac{dT_j^h}{dA} = \frac{1}{k_j}\frac{dk_j}{dA}, \quad j = 1, 2. \tag{3.4.6}$$

By (3.1.1), (3.2.5), $R_j = \eta K_j / \lambda L_j$ and $K_j = k_j N_j$

$$\frac{1}{R_j}\frac{dR_j}{dA} = \frac{1}{F_j}\frac{dF_j}{dA} = \frac{1}{K_j}\frac{dK_j}{dA} = \frac{1}{k_j}\frac{dk_j}{dA} + \frac{1}{N_j}\frac{dN_j}{dA}, \quad j = 1, 2. \tag{3.4.7}$$

3.5 Concluding Remarks

This chapter proposed a two-regional model with endogenous savings, consumption, time allocation. We built the regional model to explain endogenous accumulation of capital and interregional distribution of capital in a perfectly competitive economy. We showed how the regional differences in human capital levels and preferences for consumption, time allocation and wealth accumulation determine the regional trade patterns and regional differences in living conditions. We also analyzed the effects of changes in the human capital level and regional amenity on the economy.

4 Regional Sexual Division of Labor and Economic Growth

Over the years there have been a number of attempts to modify neoclassical consumer theory to deal with economic issues about endogenous labor supply, family structure, working hours and the valuation of traveling time (Becker, 1965, 1976, Mills and Hamilton, 1985, Folbre, 1986, Chiappori, 1988, 1992, Persson and Jonung, 1997, 1998). For instance, there is an increasing amount of economic literature about sexual division of labor, marriage and divorce, and decision-makings about family size. It has been argued that the increasing returns from human capital accumulation represent a powerful force creating a division of labor in the allocation of time between the male and female population (e.g., Becker, 1985). There are also studies on the relationship between economic growth and the family distribution of income (e.g., Fei, Ranis and Kuo, 1978). There are studies of the female labor supply. Women choose levels of market time on the basis of wage rates and incomes. Lifetime variations in costs and opportunities - due to children, unemployment of the spouse, and general business cycle variations - influence the timing of female labor participation (e.g., Mincer, 1962, Smith, 1977, Heckman and Macurdy, 1980). Possible sexual discrimination in labor markets has also attracted the attention from economists (e.g., Cain, 1986, Lazear and Rosen, 1990). The gains from marriage may be reduced as people become rich and educated. The growth in the female population's earning power may raise the forgone value of their time spent at child care, education and other household activities, which may reduce the demand for children and encourage a substitution away from parental activities. Divorce rates, fertility, and labor participation rates may interact in much more complicated ways. Decision making about on family size is extremely complicated (e.g., Becker, 1981, Weiss and Willis, 1985).

Although there is an increasing literature on dynamics of sexual division of labor and consumption in theoretical economics, interactions between regional economic growth and sexual division of labor and consumption are rarely modeled in an interregional context. The purpose of this chapter is to propose a simple two-region growth model of sexual division of labor with endogenous economic growth. This chapter is organized as follows. Section 4.1 defines the basic model. Section 4.2 provides the conditions for the existence of regional economic equilibria. Section 4.3 analyzes the impact of changes in the regional amenity levels on the equilibrium structure. Section 4.4 discusses possible directions of extending the model. This chapter is based on Zhang (1998b, 1995e).

4.1 Regional Growth with Sexual Division of Labor

The model under consideration is similar to the two-region model proposed in Chapter 2, except that we extend the model to include endogenous sexual division of labor. The system consists of male and female groups of people. Each family consists of four members - father, mother, son and daughter. The number, N, of families is fixed. The population of the economy is equal to $4N$. There is division of labor in the family. The children consume goods and accumulate knowledge through education. The parents have to do home work and find job for family's living. The father and mother may either do home work or do business. The working time of the father and the mother may be different. It is assumed that working time of the two adults are determined by maximizing the family's utility function subject to the family and the available time constraints. Possibility of divorce is omitted. It is assumed that the young people get educated before they get married and join labor market. It is assumed that the husband and wife pass away at the same time. When the parents pass away, the son and the daughter respectively find their marriage partner and get married. The property left by the parents is shared equally by the two children. The children are educated so that they have the same human capital and preference structure as their parents when they get married. When a new family is formed, the young couple join the labor market and have two children. As all the families are identical, the family structure is invariant over time under these assumptions.

A family is free to choose the region where the family works and lives. We assume that the four members of a family live and work in the same region. Each region has fixed land. It is assumed that land quality, climates and environment are homogenous within each region, but they may be different between the two regions. The land is used only for housing. We select commodity to serve as numeraire, with all the other prices being measured relative to its price. We neglect transportation cost of commodities between and within regions.

Only one commodity is produced and the commodity is composed of homogeneous qualities. It is produced by combining two factors of production - labor and capital. The commodity is either invested or consumed and is selected to serve as numeriare. Markets are characterized by perfect competition. It is assumed that labor and capital are always fully employed. To further describe the model, we introduce the following variables

L_j — the given territory size of region j, $j = 1, 2$;

$K(t)$ — the level of total capital stocks at time t;

$N_j(t)$ — the number of families in region j at time j, $j = 1, 2$;

$K_j(t)$ — the level of capital stocks employed by region j;

$F_j(t)$ — the output of region j;

$y_j(t)$ — the net income per family in region j;

$k_j(t)$ — the level of capital stocks owned by region j 's per family at time t;

$c_j(t)$ — the consumption level of per family in region j;

$s_j(t)$ — the savings made by per family in region j;

$l_j(t)$ — the lot size per family in region j;

$T_{1j}(t)$ and $T_{2j}(t)$ — the working time of the husband and the wife in region j, respectively;

$T_{1j}^h(t)$ and $T_{2j}^h(t)$ — the leisure time (or time-at-home) of the husband and the wife in region j, respectively; and

$w_{1j}(t)$ and $w_{2j}(t)$ — the wage rates per unity of working time of the husband and the wife in region j;

$R_j(t)$ and $r(t)$ — the land rent of region j and the rate of interest.

The production sectors

The qualified labor force, $N_j^*(t)$, employed by region j is given as follows

$$N_j^*(t) = z_{1j}T_{1j}N_j + z_{2j}T_{2j}N_j, \quad j = 1, 2 \tag{4.1.1}$$

where z_{1j} and z_{2j} are the human capital index of region j 's male and female population, respectively. We assume z_{1j} and z_{2j} to be constant.

The production functions of the two regions are specified as

$$F_j(t) = K_j^\alpha N_j^{*\beta}, \quad \alpha, \beta > 0, \quad \alpha + \beta = 1, \quad j = 1, 2. \tag{4.1.2}$$

The marginal conditions are given by

$$r = \frac{\alpha F_j}{K_j}, \quad w_{mj} = \frac{\beta z_{mj} F_j}{N_j^*}, \quad m, j = 1, 2. \tag{4.1.3}$$

The rate of interest is identical in the whole economy and the wage rates, w_{mj}, may be different. As $w_{1j} / w_{2j} = z_j \equiv z_{1j} / z_{2j}$, the wage ratio between the husband and wife is always equal to the ratio of their working efficiency levels.

Behavior of households

Each family gets income from land ownership, capital ownership and wages. In order to define net incomes, it is necessary to determine land ownership structure. We specify land revenue as follows. We assume that each family owns L_1 / N amount of land in region 1 and L_2 / N in region 2 and it is impossible to sell land but it is free to rent one's own land to others. Under this assumption, the land revenue, R_0, per family is given as follows

$$R_0(t) = \frac{L_1 R_1(t) + L_2 R_2(t)}{N}. \tag{4.1.4}$$

This assumption implies that a worker receives the land revenue, R_0, irrespective of whether his dwelling site is located in region 1 or region 2.

If we denote $k_j(t)$ capital stocks owned by per family in region j, the interest payment per family is given by $rk_j(t)$. Under the above land ownership distribution, the net income per family, $y_j(t)$, in region j consists of the wage income, $T_{1j}(t)w_{1j}(t) + T_{2j}(t)w_{2j}(t)$, land revenue, R_0, and interest payment, $rk_j(t)$, i.e.

$$y_j(t) = T_{1j}(t)w_{1j}(t) + T_{2j}(t)w_{2j}(t) + r(t)k_j(t) + R_0(t), \quad j = 1, 2. \tag{4.1.5}$$

It is assumed that a typical family's utility level, $U_j(t)$, in region j is dependent on the husband's and wife's person's leisure time, $T_{1j}^h(t)$ and $T_{2j}^h(t)$, lot size, $l_j(t)$, consumption level, $c_j(t)$, of community, and the level of the family's wealth, $k_j(t) + s_j(t) - \delta_k k_j(t)$, where δ_k is the fixed depreciation rate of capital. The utility functions are specified as follows

$$U_j(t) = A_j T_{1j}^h(t)^{\sigma_1} T_{2j}^h(t)^{\sigma_2} l_j(t)^{\eta} c_j(t)^{\xi} \{k_j(t) + s_j(t) - \delta_k k_j(t)\}^{\lambda}, \quad \sigma_1, \sigma_2, \eta, \xi, \lambda > 0, \; \sigma_1 + \sigma_2 + \eta + \xi + \lambda = 1 \tag{4.1.6}$$

in which σ_1, σ_2, η, ξ, and λ are respectively the household's propensity to use the husband's leisure time, the wife's leisure time, to utilize lot size, to consume the commodity, and to hold wealth. In (4.1.6), A_j is region j's amenity level.

Let T_0 denote the total available time. The time constraint requires that the amounts of time allocated to each specific use add up to the time available

$$T_{mj} + T_{mj}^h = T_0, \quad m, j = 1, 2. \tag{4.1.7}$$

The budget constraints are given by

$$l_j R_j + c_j + s_j = y_j, \quad j = 1, 2. \tag{4.1.8}$$

By (4.1.4) and (4.1.6), we may rewrite (4.1.7) as follows

$$l_j R_j + c_j + s_j + T_{1j} w_{1j} + T_{2j} w_{2j} = R_0 + T_{1j}^h w_j + T_{2j}^h w_j + rk_j, \quad j = 1, 2. \tag{4.1.9}$$

A typical person maximizes the person's utility subject to (4.1.9). The optimal problems have the following solutions

$$T_{1j}^h w_{1j} = \sigma_1 \Omega_j, \; T_{2j}^h w_{2j} = \sigma_2 \Omega_j, \; l_j R_j = \eta \Omega_j, \quad c_j = \xi \Omega_j, \quad s_j = \lambda \Omega_j - \delta k_j, \quad j = 1, 2 \tag{4.1.10}$$

where

$$\Omega_j \equiv R_0 + T_0 w_{1j} + T_0 w_{2j} + rk_j + \delta k_j, \quad \delta \equiv 1 - \delta_k.$$

By (4.1.10) and $w_{1j} / w_{2j} = z_j$, we get

$$\frac{T_{1j}^h}{T_{2j}^h} = z_j \frac{\sigma_1}{\sigma_2}.$$

The ratio of husband's and wife's time at home is positively related to the ratio of the marginal values of the husband and wife's time at home and negatively related to the ratio of the husband's and wife's working efficiency levels.

According to the definitions of k_j and s_j, the capital accumulation of the family in region j is given by

$$\frac{dk_j}{dt} = s_j - \delta_k k_j .$$

Substituting s_j in (4.1.10) into these two equations yields

$$\frac{dk_j}{dt} = \lambda\Omega_j - k_j , \quad j = 1, 2. \tag{4.1.11}$$

As families are assumed to be freely mobile between the two regions, the utility level of families should be equal, irrespective of in which region they live, i.e.

$$U_1(t) = U_2(t). \tag{4.1.12}$$

Balance of demand and supply

By the definitions of K, k_j and N_j

$$K = k_1 N_1 + k_2 N_2 . \tag{4.1.13}$$

The assumption that labor force, capital stocks and land are fully employed is represented by

$$N_1 + N_2 = N, \quad K_1 + K_2 = K, \quad l_j N_j = L_j , \quad j = 1, 2. \tag{4.1.14}$$

The balance of demand for and supply of industrial commodity is given by

$$(c_1 + s_1) N_1 + (c_2 + s_2) N_2 = F_1 + F_2 . \tag{4.1.15}$$

We have thus built the model. The system has 35 variables, N_j, K_j, F_j, c_j, s_j, k_j, l_j, T_{1j}, T_{1j}^h, T_{2j}, T_{2j}^h, y_j, U_j, w_{1j}, w_{2j}, R_j $(j = 1, 2)$, K, r and R_0. We now examine conditions for the existence of equilibria of the dynamic system.

4.2 Equilibrium Structure of Regional Competition

This section is concerned with the existence of economic equilibria. By (4.1.11)

$$\lambda\Omega_j = k_j, \quad j = 1, 2. \tag{4.2.1}$$

Substituting (4.2.1) into (4.1.10) yields

$$T_{1j}^h w_{1j} = \frac{\sigma_1 k_j}{\lambda}, \quad T_{2j}^h w_j = \frac{\sigma_2 k_j}{\lambda}, \quad l_j R_j = \frac{\eta k_j}{\lambda}, \quad c_j = \frac{\xi k_j}{\lambda},$$
$$s_j = \delta_k k_j, \quad j = 1, 2. \tag{4.2.2}$$

Substituting (4.2.2) into (4.1.15) yields

$$\delta_1 K = F_1 + F_2 \tag{4.2.3}$$

where

$$\delta_1 \equiv \frac{\xi}{\lambda} + \delta_k .$$

By $r = \alpha F_1 / K_1 = \alpha F_2 / K_2$ in (4.1.2) and (4.2.3), we have

$$r = \alpha \delta_1 \tag{4.2.4}$$

where we use $K_1 + K_2 = K$. By (4.1.1), $r = \alpha F_1 / K_1 = \alpha F_2 / K_2$ in (4.1.2) and (4.2.4), we get

$$N_j^* = \delta_1^{1/\beta} K_j, \quad j = 1, 2. \tag{4.2.5}$$

By (4.2.5) and (4.1.2)

$$w_{mj} = \frac{\beta z_{mj}}{\delta_1^{\alpha/\beta}}, \quad m, j = 1, 2. \tag{4.2.6}$$

Substituting $l_j = L_j / N_j$ and (4.2.2) into (4.1.6) and then using (4.1.12), we obtain

$$\frac{k_1}{k_2} = A\Lambda^{\eta\upsilon} \tag{4.2.7}$$

where we use (4.2.6) and

$$\Lambda \equiv \frac{N_1}{N_2}, \quad \upsilon \equiv \frac{1}{1-\eta}, \quad A \equiv \left(\frac{z_{11}}{z_{12}}\right)^{\sigma_1\upsilon} \left(\frac{z_{21}}{z_{22}}\right)^{\sigma_2\upsilon} \left(\frac{A_2}{A_1}\right)^{\upsilon} \left(\frac{L_2}{L_1}\right)^{\upsilon\eta}.$$

By the definition of Λ and $N_1 + N_2 = N$, we have

$$N_1 = \frac{\Lambda N}{1+\Lambda}, \quad N_2 = \frac{N}{1+\Lambda}. \tag{4.2.8}$$

By $l_j R_j = \eta k_j / \lambda$ in (4.2.2), $l_j = L_j / N_j$ and (4.1.4), we get

$$R_0 = \frac{\eta K}{\lambda N} \tag{4.2.9}$$

where we use (4.1.13). By the definitions of Ω_j and (4.2.1), we get

$$R_0 + w_j = \delta_2 k_j, \quad j = 1, 2 \tag{4.2.10}$$

where $w_j \equiv (w_{1j} + w_{2j})T_0$ and

$$\delta_2 \equiv \frac{1}{\lambda} - \delta - r = \frac{\sigma_1 + \sigma_2 + \eta + \beta\xi}{\lambda} + \beta\delta_k > 0.$$

By (4.2.10), we have

$$k_1 - k_2 = w_0 \equiv (z-1)\frac{w_2}{\delta_2} > 0 \tag{4.2.11}$$

where

$$z \equiv \frac{z_{11} + z_{21}}{z_{12} + z_{22}}.$$

In the remainder of this chapter, we assume $z > 1$ i.e.

$$z_{11} + z_{21} > z_{12} + z_{22}.$$

The requirement means that the sum of the husband's and wife's working efficiency in region 1 is higher than that in region 2. By (4.2.11) we see that the level of capital stocks owned by per family in region 1 is higher than that in region 2, i.e., $k_1 > k_2$. By (4.2.7) and (4.2.11), we solve

$$k_1 = \frac{w_0 A\Lambda^{\eta\upsilon}}{A\Lambda^{\eta\upsilon} - 1}, \quad k_2 = \frac{w_0}{A\Lambda^{\eta\upsilon} - 1}. \tag{4.2.12}$$

It is necessary to require: $A\Lambda^{\eta\upsilon} > 1$. Substituting (4.2.9) and k_2 in (4.2.12) into $R_0 + w_2 = \delta_2 k_2$ in (4.2.10), we get

$$K = \frac{\lambda N T_0 w_2}{\eta\left(A\Lambda^{\eta\upsilon} - 1\right)}\left(z - A\Lambda^{\eta\upsilon}\right). \tag{4.2.13}$$

Since $A\Lambda^{\eta\upsilon} > 1$, for $K > 0$ it is necessary to require: $z > A\Lambda^{\eta\upsilon}$. Since $k_j > 0$ are guaranteed by $A\Lambda^{\eta\upsilon} > 1$, by $T_{mj}^h w_{mj} = \sigma_m k_j / \lambda$ in (4.2.2), we see that $T_0 > T_{mj} > 0$ are satisfied if $T_0 = \sigma_m k_j / \lambda w_{mj}$, $m, j = 1, 2$. By (4.2.12), we see that these four inequalities are satisfied if

$$A\Lambda^{\eta\upsilon} > \max\left\{\frac{\lambda\delta_2 z}{\lambda\delta_2 z - \sigma_m(z - 1)}, \ \frac{\sigma_m}{\lambda\delta_2}(z - 1) + 1, \ m = 1, 2\right\}.$$

Since

$$z > \frac{\sigma_m}{\lambda\delta_2}(z - 1) + 1 > \frac{\lambda\delta_2 z}{\lambda\delta_2 z - \sigma_m(z - 1)} > 1, \quad m = 1, 2$$

we require Λ to satisfy

$$z > A\Lambda^{\eta\upsilon} > \frac{1}{\lambda\delta_2}(z - 1)\max\{\sigma_1, \sigma_2\} + 1.$$

If we require $\sigma_2 > \sigma_1$, then the above inequalities become

$$z > A\Lambda^{\eta\upsilon} > \frac{\sigma_2}{\lambda\delta_2}(z-1)+1. \tag{4.2.14}$$

Under (4.2.14), we have $K > 0$, $k_j > 0$ and $T_0 > T_{mj}^h > 0$.

Substituting (4.2.8) and (4.2.12) into (4.1.13) and then using (4.2.13), we get

$$\Phi(\Lambda) \equiv \left(A\Lambda^{\eta}+1\right)(z-1) - \left(z - A\Lambda^{\eta\upsilon}\right)(1+\Lambda)\frac{\lambda\delta_2}{\eta}. \tag{4.2.15}$$

The function, $\Phi(\Lambda)$, has the following properties

$$\begin{aligned}
\Phi &= \frac{\sigma_1 + \beta\xi + \beta\lambda\delta_k - (z-1)\sigma_2\eta/\lambda\delta_2}{\eta}(1+\Lambda)(z-1) \\
&- \frac{\sigma_2}{\lambda\delta_2}(z-1)^2, \quad at\ A\Lambda^{\eta\upsilon} = \frac{\sigma_2}{\lambda\delta_2}(z-1)+1, \\
\Phi &= (1+z\Lambda)(z-1) \quad at\ A\Lambda^{\eta\upsilon} = z, \\
\Phi^* &\equiv \frac{d\Phi}{d\Lambda} = \upsilon A\Lambda^{\eta\upsilon}(z-1) + \upsilon\lambda\delta_2 A\Lambda^{\eta\upsilon-1}(1+\Lambda) - \frac{z\lambda\delta_2}{\eta} \\
&+ \frac{\lambda\delta_2 A\Lambda^{\eta\upsilon}}{\eta} > \left\{\frac{\sigma_2\upsilon}{\lambda\delta_2}(z-1) + \upsilon + \frac{\sigma_2\upsilon}{\eta} - \frac{\lambda\delta_2}{\eta}\right\}(z-1) \\
&+ \upsilon\lambda\delta_2, \quad \text{for } z > A\Lambda^{\eta\upsilon} > \frac{\sigma_2}{\lambda\delta_2}(z-1)+1.
\end{aligned} \tag{4.2.16}$$

Since $\Phi > 0$ at $A\Lambda^{\eta\upsilon} = z$, we see that the equation, $\Phi(\Lambda) = 0$, has at least one solution satisfying (4.2.14) if $\Phi < 0$ at $A\Lambda^{\eta\upsilon} = (z-1)\sigma/\lambda\delta_2 + 1$. By the first equation in (4.2.16), we see that $\Phi < 0$ at $A\Lambda^{\eta\upsilon} = (z-1)\sigma/\lambda\delta_2 + 1$ is held if

$$\sigma_1 + \beta\xi + \beta\lambda\delta_k > (z-1)\frac{\sigma_2\eta}{\lambda\delta_2}.$$

As $\sigma_2 \eta / \lambda \delta_2 < \min\{\sigma_2, \eta\}$, if $z - 1$ is small, then the requirement is satisfied. If $\Phi^* > 0$ under (4.2.14), then the equation has a unique solution in the interval. By (4.2.16), we see that it is acceptable to require $\Phi^* > 0$. For instance, if we have

$$\eta + (z - 1)\frac{\sigma_2 + \eta}{\lambda \delta_2} > \frac{z - 1}{\upsilon}$$

then $\Phi^* > 0$ is held under (4.2.14).

Proposition 4.2.1.
We assume

$$\sigma_1 + \beta \xi + \beta \lambda \delta_k > (z - 1)\frac{\sigma_2 \eta}{\lambda \delta_2}, \quad \eta + (z - 1)\frac{\sigma_2 + \eta}{\lambda \delta_2} > \frac{z - 1}{\upsilon}.$$

The dynamic system has a unique equilibrium. The unique equilibrium values of the variables are given by the following procedure: r by (4.2.4) $\rightarrow$ w_{mj}, $m, j = 1, 2$ by (4.2.6) $\rightarrow$ Λ by (4.2.15) $\rightarrow$ N_j by (4.2.8) $\rightarrow$ k_j by (4.2.12) $\rightarrow$ K by (4.2.13) $\rightarrow$ R_0 by (4.2.9) $\rightarrow$ T_{mj}^h by (4.2.2) $\rightarrow$ $T_{mj} = T_0 - T_{mj}^h$ $\rightarrow$ N_j^* by (4.1.1) $\rightarrow$ K_j by (4.2.5) $\rightarrow$ F_j by (4.1.2) $\rightarrow$ $l_j = L_j / N_j$ $\rightarrow$ R_j, c_j and s_j by (4.2.2) $\rightarrow$ y_j by (4.1.5) $\rightarrow$ U_j by (4.1.6).

The assumptions in Proposition 4.2.1 are satisfied if the difference between the two regions' levels of working efficiency is appropriately small. It should be remarked that the dynamic system may have a unique equilibrium even if the requirements in Proposition 4.2.1 are not satisfied. In the remainder of this chapter, we examine effects of changes in some parameters. We accept the assumptions in Proposition 4.2.1 in the remainder of this chapter.

4.3 Regional Amenity and the Equilibrium Structure

This section is concerned with the effects of changes in the amenity levels of the two regions. By (4.2.15) and (4.2.8)

$$\Phi^* \frac{d\Lambda}{dA} = (1 - z)\Lambda^{\upsilon} - \frac{\lambda \delta_2 \Lambda^{\eta \upsilon}}{\eta}(1 + \Lambda) < 0,$$

$$\frac{1}{N_1}\frac{dN_1}{dA} = \frac{1}{(1+\Lambda)\Lambda}\frac{d\Lambda}{dA} < 0, \quad \frac{1}{N_2}\frac{dN_2}{dA} = -\frac{1}{1+\Lambda}\frac{d\Lambda}{dA} > 0. \tag{4.3.1}$$

By the definition of A, we see that a change in A may be caused by changes in the working efficiency, the propensities, and the amenity levels. In this section, we require that a change in A is caused by changes in A_2 or A_1. Hence, an increase in A may be interpreted as an improvement in region 2's amenity level or a decrease in region 1's amenity level. As region 2's amenity is improved, some of region 1's population will migrate to region 2.

By $y_j = (\xi/\lambda + \eta/\lambda + \delta_k)k_j$ and (4.2.12), we have

$$\frac{1}{y_1}\frac{dy_1}{dA} = \frac{1}{k_1}\frac{dk_1}{dA} = -\frac{1}{(A\Lambda^{\eta\upsilon}-1)A} - \frac{\eta\upsilon A\Lambda^{\eta\upsilon-1}}{A\Lambda^{\eta\upsilon}-1}\frac{d\Lambda}{dA},$$
$$\frac{1}{y_2}\frac{dy_2}{dA} = \frac{1}{k_2}\frac{dk_2}{dA} = -\frac{\Lambda^{\eta\upsilon}}{A\Lambda^{\eta\upsilon}-1} - \frac{\eta\upsilon A\Lambda^{\eta\upsilon-1}}{A\Lambda^{\eta\upsilon}-1}\frac{d\Lambda}{dA}. \tag{4.3.2}$$

It is difficult to explicitly judge the signs of dk_j/dA and dy_j/dA. By (4.2.4) and (4.2.6)

$$\frac{dr}{dA} = \frac{dw_{mj}}{dA} = 0, \quad m, j = 1, 2. \tag{4.3.3}$$

The rate of interest and the two regions' wage rates are not affected. By (4.2.9) and (4.2.10), we get

$$\frac{1}{R_0}\frac{dR_0}{dA} = \frac{1}{K}\frac{dK}{dA} = \frac{\lambda\delta_2 N}{\eta K}\frac{dk_2}{dA}. \tag{4.3.4}$$

The per family's revenue, R_0, from land ownership and the total capital stocks, K, of the economy may be either increased or reduced. By $l_j = L_j/N_j$

$$\frac{1}{l_1}\frac{dl_1}{dA} = -\frac{1}{N_1}\frac{dN_1}{dA} > 0, \quad \frac{1}{l_2}\frac{dl_2}{dA} = -\frac{1}{N_2}\frac{dN_2}{dA} < 0. \tag{4.3.5}$$

The lot size per family in region 1 (4.1) is increased (reduced). By (4.2.2) and $T_{mj} + T_{mj}^h = T_0$, we get

$$\frac{1}{c_j}\frac{dc_j}{dA} = \frac{1}{k_j}\frac{dk_j}{dA}, \quad \frac{dT_{mj}}{dA} = -\frac{1}{k_j}\frac{dT_{mj}^h}{dA},$$
$$\frac{1}{T_{mj}^h}\frac{dT_{mj}^h}{dA} = \frac{1}{k_j}\frac{dk_j}{dA}, \quad m, j = 1, 2. \tag{4.3.6}$$

By (4.1.1), (4.2.5), $R_j = \eta K_j / \lambda L_j$, and $K_j = k_j N_j$, we have

$$\frac{1}{R_j}\frac{dR_j}{dA} = \frac{1}{F_j}\frac{dF_j}{dA} = \frac{1}{K_j}\frac{dK_j}{dA} = \frac{1}{k_j}\frac{dk_j}{dA} + \frac{1}{N_j}\frac{dN_j}{dA}, \quad j = 1, 2. \tag{4.3.7}$$

4.4. On Complex of Sexual Division of Labor and Consumption

This section was regional economic evolution with sexual division of labor. Section 4.1 synthesized the growth models with capital and time distribution into a single framework. This chapter examined sexual division of labor within the one-sector framework with endogenous capital accumulation, capital distribution between regions, and time distribution between work and leisure. We were mainly concerned with behavior of the dynamic system when the female population takes part in labor market. We also examined the effects of regional amenity on the equilibrium economic structure and sexual division of labor.

The model in this chapter may be extended in multiple ways. We may further examine the behavior of the model. For instance, we may examine effects of changes in some other parameters such as the population and the knowledge utilization efficiency on economic growth and the sexual division of labor. From a long-run development point of view, it is reasonable to treat the preference parameters as endogenous variables. We may also extend the production and knowledge accumulation to include multiple sectors and R&D activities. The two sexes may obtain different amenities from different professions. This chapter does not specify how the family game is actually played. In this chapter, improvement in the female population's human capital is assumed to be costless and have no impact on the male population's human capital. We assumed that each man or woman has the ideal family as well as an identical family structure. Some people may actually remain unmarried. A couple may divorce, which implies no-existence of any family utility function. Decisions about family size as a life style, the above conclusion may not valid even if we assume that there is not discrimination (Zhang, 1997d).

5 A Two-group Regional Growth Model

People's preferences and human capital are different. Heterogeneity in preferences among individuals is an important determinant of spatial structure. Individuals may differentiate quite strongly among particular regions based on differences in local climate or amenities available. These differences may have impact on the location of consumers and producers. They may also create spatial agglomeration economies in consumption and production. Many economic theories which explain various aspects of the complexity of economic geography have been proposed. For instance, economic growth theories emphasize economic dynamics with capital, population and knowledge accumulation, but neglect spatial characteristics of economic activities. Some spatial economic models take account of endogenous capital and population growth, but neglect interregional issues (e.g., Henderson, 1985, Miyao, 1987, Richardson, 1977). This chapter tries to propose a dynamic economic growth model with interregional trade, synthesizing the ideas in the standard one-sector neoclassical growth model and the Kaldor-Pasinetti two-group model.

The previous chapters extended the one-sector neoclassical growth model into an interregional context with time distribution and sexual division of labor and consumption. Nevertheless, we assumed that the population is homogenous. This is obviously a strict assumption. The purpose of this chapter is to deal with growth of interregional economies with heterogenous population, based on the Kaldor-Pasinetti model. The significance of the Kaldor-Pasinetti two-class model lies in that it explicitly takes account of endogenous wealth and income distribution among various social and economic groups (see, for instance, Sato, 1966, Pasinetti, 1974, Salvadori, 1991, Panico and Salvadori, 1993). But this approach completely neglects location issues. Although the urban economist has proposed multi-group models (e.g., Yellin, 1974, Yinger, 1976, Zhang, 1989, 1993a), these models are basically developed within static frameworks with fixed incomes. These models have not addressed dynamics of wealth and income distribution among various groups and their relative residential location. Our model is developed on the basis of these models.

The remainder of this chapter is organized as follows. Section 5.1 defines the basic two-group model of capital accumulation and regional location. Section 5.2 analyzes conditions for the existence of equilibria. Section 5.3 illustrates possible extensions of the model. The appendix examines the general case of the model. This chapter is based on Zhang (1995b).

5.1 The Regional Model with Multiple Groups

We consider an economic system which consists of two regions, indexed by 1 and 2, respectively. Each region has a single production sector. Production is carried out by combining capital and labor. It is assumed that the two regions' product is qualitatively homogenous and is either consumed or invested. We assume that the population is classified into two groups, indexed by 1 and 2, respectively. A person from any group is free to choose his residential location. We assume that any person chooses the same region where he works and lives. Each region has fixed land. It is assumed that land quality, climates and environment are homogenous within each region, but climates and environment may be different between the two regions. The land is used only for housing. We select commodity to serve as numeraire, with all the other prices being measured relative to its price. We neglect transportation cost of commodities between and within regions. Although it is conceptually not difficult to introduce transportation cost function and to provide balance conditions for demand and supply and for price equalization conditions with transportation cost, the problem will become analytically too complicated. The assumption of zero transportation cost of commodities implies price equality for the commodity between the two regions. As the amenity levels are fixed and land is immobile, wage rates and land rent may not be equal between the two regions.

We introduce

N_m —the given population of group m, $m = 1, 2$;

L_j —the given territory size of region j, $j = 1, 2$;

$K(t)$ —the total capital stocks of the economy at time t;

$F_j(t)$ —the output levels of region j 's industrial sector at time t;

$K_j(t)$ —the levels of capital stocks employed by region j 's production sector;

$N_{mj}(t)$ — the number of group m 's labor force employed by region j 's production sector, $m, j = 1, 2$;

$c_{mj}(t)$ and $s_{mj}(t)$ — per capita's consumption level of commodity and savings made by per capita of group m in region j;

$l_{mj}(t)$ —the lot size per capita of group m in region j;

$y_{mj}(t)$ —the net income per capita of group m in region j;

$r(t)$ —the rate of interest; and

$w_{mj}(t)$ —group m 's wage rate in region j; and

$R_j(t)$ —region j 's land rent.

We now describe the model.

The production sectors

We define region j 's labor input, $N_j^*(t)$, as follows

$$N_j^*(t) = z_{1j}N_{1j} + z_{2j}N_{2j}, \quad j = 1, 2 \tag{5.1.1}$$

in which z_{mj} are the working efficiency index of group m 's labor force in region j. For simplicity, we assume z_{mj} to be constant. For convenience of discussion, we always assume: $z_{m1} - z_{m2} \neq 0$. That is, workers from the same group are different in working efficiency in different regions. The case of $z_{m1} = z_{m2}$ can be similarly discussed.

The production functions of the two regions are specified as follows

$$F_j(t) = K_j^\alpha N_j^{*\beta}, \quad \alpha, \beta > 0, \quad \alpha + \beta = 1, \quad j = 1, 2. \tag{5.1.2}$$

The marginal conditions are given by

$$r = \frac{\alpha F_j}{K_j}, \quad w_{mj} = \frac{\beta z_{mj} F_j}{N_j^*}, \quad m, j = 1, 2. \tag{5.1.3}$$

The rate of interest is identical in the whole economy and the wage rates, w_{mj}, may be different between the two regions.

Behavior of households

In order to define net incomes, it is necessary to determine land ownership structure. To simplify the discussion, we specify the land revenue as follows. Let N denote the total population of the economy, i.e., $N = N_1 + N_2$. We assume that each person owns L_1 / N amount of land in region 1 and L_2 / N in region 2 and it is impossible to sell land but it is free to rent one's own land to others. Under this assumption, the land revenue, R_0, per person is given as follows

$$R_0(t) = \frac{L_1 R_1(t) + L_2 R_2(t)}{N}. \tag{5.1.4}$$

This assumption implies that any person in the system receives the land revenue, R_0, irrespective of whether his dwelling site is located in region 1 or region 2.

If we denote $k_{mj}(t)$ capital stocks owned by per capita of group m in region j, the interest payment per capita of group m in region j is given by $rk_{mj}(t)$. Under the above specified land ownership distribution, the net income, $y_{mj}(t)$, per capita of group m in region j consists of the wage income, $w_{mj}(t)$, land revenue, R_0, and interest payment, $rk_{mj}(t)$, i.e.

$$y_{mj}(t) = w_{mj}(t) + r(t)k_{mj}(t) + R_0(t), \quad j = 1, 2. \tag{5.1.5}$$

It is assumed that the utility level, $U_{mj}(t)$, that a typical person of group m obtains in region j is dependent on the lot size, $l_{mj}(t)$, the person's consumption level, $c_{mj}(t)$, of community, and the level of the person's wealth, $k_{mj} + s_{mj} - \delta_k k_{mj}$, where δ_k is the fixed depreciation rate of capital. The utility functions are specified as follows

$$U_{mj}(t) = A_{mj} l_{mj}(t)^{\eta_m} c_{mj}(t)^{\xi_m} \{k_{mj}(t) + s_{mj}(t) - \delta_k k_{mj}(t)\}^{\lambda_m},$$
$$\eta_m, \xi_m, \lambda_m > 0, \quad \eta_m + \xi_m + \lambda_m = 1 \tag{5.1.6}$$

in which η_m, ξ_m, and λ_m are respectively group m's propensity to use lot size, to consume commodity, and to hold wealth. In (5.1.6), A_{mj} is called region j's amenity level for group m. In this chapter, amenities are considered as locational parameters.

The budget constraints are given by

$$l_{mj} R_j + c_{mj} + s_{mj} = y_{mj}, \quad m, j = 1, 2. \tag{5.1.7}$$

A typical person maximizes the person's utility subject to its budget constraint. The optimal problems have the following solutions

$$l_{mj} R_j = \eta_m \Omega_{mj}, \quad c_{mj} = \xi_m \Omega_{mj}, \quad s_{mj} = \lambda_m \Omega_{mj} - \delta_k k_{mj} \tag{5.1.8}$$

where

$$\Omega_{mj} \equiv y_{mj} + \delta k_{mj}, \quad \delta \equiv 1 - \delta_k .$$

According to the definitions of k_{mj} and s_{mj}, the capital accumulation of a typical person of group m in region j is given by

$$\frac{dk_{mj}}{dt} = s_{mj} - \delta_k k_{mj} .$$

Substituting s_{mj} in (5.1.8) into these four equations yields

$$\frac{dk_{mj}}{dt} = \lambda_m \Omega_{mj} - k_{mj}, \quad m, \, j = 1, \, 2 . \tag{5.1.9}$$

As people are freely mobile between the two regions, the utility level of the same group are equal, irrespective of in which region they live, i.e.

$$U_{m1}(t) = U_{m2}(t), \quad m = 1, \, 2 . \tag{5.1.10}$$

Balance of demand and supply
By the definitions of K, k_{mj} and N_{mj}

$$K = \sum\nolimits_{m,j} k_{mj} N_{mj} . \tag{5.1.11}$$

The assumption that labor force, capital stocks and land are fully employed is represented by

$$N_{m1} + N_{m2} = N_m, \quad K_1 + K_2 = K, \quad l_{1j} N_{1j} + l_{2j} N_{2j} = L_j, \quad m, \, j = 1, \, 2 . \tag{5.1.12}$$

The balance of demand for and supply of industrial commodity is given by

$$\sum\nolimits_{m,j} \left(c_{mj} + s_{mj} \right) N_{mj} = F_1 + F_2 . \tag{5.1.13}$$

We thus built the model. The system has 39 variables, c_{mj}, s_{mj}, k_{mj}, l_{mj}, w_{mj}, y_{mj}, U_{mj} ($m, \, j = 1, \, 2$), N_j^*, K_j, F_j, R_j ($j = 1, \, 2$), K, r and R_0. We now examine conditions for the existence of equilibria of the dynamic system.

5.2 The Equilibrium Regional Structure

This section is concerned with conditions for the existence of equilibria of the dynamic system when the two groups' propensities are equal, i.e.

$$\eta \equiv \eta_1 = \eta_2, \quad \xi \equiv \xi_1 = \xi_2, \quad \lambda \equiv \lambda_1 = \lambda_2 .$$

The two groups are different in the working efficiency and amenity levels they obtain in different regions. The appendix provides the conditions for the existence of economic equilibria when the two groups have different propensities.

By (5.1.9), at equilibrium

$$\lambda_m \Omega_{mj} = k_{mj}, \quad m, j = 1, 2. \tag{5.2.1}$$

Substituting (5.2.1) into (5.1.8) yields

$$l_{mj} R_j = \frac{\eta k_{mj}}{\lambda}, \quad c_{mj} = \frac{\xi k_{mj}}{\lambda}, \quad s_{mj} = \delta_k k_{mj}, \quad j, m = 1, 2. \tag{5.2.2}$$

Substituting (5.2.2) into (5.1.6) and then using (5.1.10), we get

$$k_m \equiv \frac{k_{m1}}{k_{m2}} = \left(\frac{A_{m2}}{A_{m1}}\right)^{\eta} R^{\eta}, \quad m = 1, 2 \tag{5.2.3}$$

where $R \equiv R_1 / R_2$. By (5.2.3), we obtain

$$k_1 = A k_2 \tag{5.2.4}$$

where $A \equiv \left(A_{12} A_{21} / A_{11} A_{22}\right)^{\eta}$.

Substituting c_{mj} and s_{mj} in (5.2.2) into (5.1.13) yields

$$F_1 + F_2 = \delta^* K \tag{5.2.5}$$

where we use (5.1.11) and $\delta^* \equiv \xi / \lambda + \delta_k$.

By (5.1.2) and $r = \alpha F_1 / K_1 = \alpha F_2 / K_2$ in (5.1.3), we have

$$\Lambda = \frac{K_1}{N_1^*} = \frac{K_2}{N_2^*}.$$

By this equation and (5.1.3)

$$r = \frac{\alpha}{\Lambda^\beta}, \quad w_{mj} = \beta z_{mj} \Lambda^\alpha, \quad m, j = 1, 2. \tag{5.2.6}$$

Substituting $l_{mj} = \eta k_{mj} / \lambda R_j$ in (5.2.2) into the land constraint equations in (5.1.12) yields

$$k_{1j} N_{1j} + k_{2j} N_{2j} = \frac{\lambda L_j R_j}{\eta}, \quad j = 1, 2. \tag{5.2.7}$$

Adding the two equations in (5.2.7), we get

$$R_0 = \frac{\eta K}{\lambda N} \tag{5.2.8}$$

where we use (5.1.4).

By $r = \alpha F_j / K_j$ in (5.1.3) and $r = \alpha / \Lambda^\beta$ in (5.2.6), $F_j = K_j / \Lambda^\beta$. Substituting $F_j = K_j / \Lambda^\beta$ into (5.2.5) yields

$$\Lambda^\beta = \frac{1}{\delta^*} \tag{5.2.9}$$

where we use $K_1 + K_2 = K$. We thus solved Λ by (5.2.9) and r and w_{mj} by (5.2.6).

By (5.2.1), (5.1.5) and the definitions of Ω_{mj}, we have

$$\upsilon k_{mj} \equiv R_0 + w_{mj}, \quad m, j = 1, 2 \tag{5.2.10}$$

where $\upsilon \equiv 1/\lambda - \delta - r > 0$. By (5.2.10)

$$k_m = \frac{R_0 + w_{m1}}{R_0 + w_{m2}}, \quad m = 1, 2.$$

Substituting these two equations into (5.2.4) yields

$$R_0^2 - 2b_1 R_0 - b_2 = 0 \tag{5.2.11}$$

where

$$b_1 \equiv \frac{Az_{21} + Az_{12} - z_{11} - z_{22}}{2(1 - A)} \beta\Lambda^{\alpha},$$

$$b_2 \equiv \frac{Az_{12}z_{21} - z_{11}z_{22}}{1 - A} \beta^2 \Lambda^{2\alpha}$$

in which we use (5.2.6). If $A = 1$, R_0 is given by

$$R_0 = \frac{z_{11}z_{22} - z_{12}z_{21}}{z_{21} + z_{12} - z_{11} - z_{22}} \beta\Lambda^{\alpha}.$$

The condition, $A = 1$, i.e., $A_{11} / A_{12} = A_{21} / A_{22}$, implies that the ratio of amenity levels that each group obtains in the two regions is equal between the two groups. In this case, R_0 is meaningful if

$$z_{11}z_{22} > (<)\ z_{12}z_{21}, \quad z_{21} + z_{12} > (<)\ z_{11} + z_{22}.$$

In the remainder of this chapter, we assume $1 > A$, i.e., $A_{11} / A_{12} > A_{21} / A_{22}$. The requirement implies that the ratio of amenity levels that group 1 obtains in region 1 and region 2 is higher than the ratio that group 2 obtains in region 1 and region 2. The equation, (5.2.11), has two solutions as follows

$$R_0 = b_1 \pm \left(b_1^2 + b_2\right)^{1/2}. \tag{5.2.12}$$

It is necessary to require R_0 to be positive. In the case of $b_2 > 0$, the equation has a single meaningful solution, $R_0 = b_1 + \left(b_1^2 + b_2\right)^{1/2}$. In the case of $b_2 < 0$ and $b_1 < 0$ (or $b_1^2 + b_2 < 0$ with $b_1 > 0$), the equation has no meaningful solution. This implies that some group may be concentrated in a single region. In the case of $b_2 < 0$, $b_1 > 0$ and $b_1^2 + b_2 > 0$, the two solutions of (5.2.11) in (5.2.12) are meaningful. As $1 > A$, the sign of b_1 is the same as that of $Az_{21} + Az_{12} - z_{11} - z_{22}$ and the sign of b_2 is the same as that of $Az_{21} / z_{22} - z_{11} / z_{12}$. The case of $b_2 > 0$, i.e., $Az_{21} / z_{22} > z_{11} / z_{12}$ (under $A_{11} / A_{12} > A_{21} / A_{22}$), means that the ratio of group 2's working efficiency in region 1 and region 2 is much higher than the ratio of group 1's working efficiency in region 1 and region 2.

In the remainder of this chapter, we are only concerned with the case of $b_2 > 0$. In this case the equation, (5.2.11), has a unique positive solution

$$R_0 = b_1 + \left(b_1^2 + b_2\right)^{1/2}.$$

For given R_0, we solve K by (5.2.8) and k_{mj} by (5.2.10). By (5.2.3) we solve R. By (5.1.4) and $R_1 = RR_2$, we solve R_j as follows

$$R_1 = \frac{R_0 R N}{L_1 R + L_2}, \quad R_2 = \frac{R_0 N}{L_1 R + L_2}. \tag{5.2.13}$$

By (5.2.7) and $N_{m1} + N_{m2} = N_m$, we solve

$$N_{11} = \frac{k_{02} N_1 + N_2 - M_2 - M_1}{k_{02} - k_{01}}, \quad N_{12} = N_1 - N_{11},$$

$$N_{21} = \frac{k_{01} M_2 + k_{02} M_1 - k_{01} k_{02} N_1 - k_{01} N_2}{k_{02} - k_{01}}, \quad N_{22} = N_2 - N_{21}$$

(5.2.14)

where

$$M_j \equiv \frac{\lambda L_j R_j}{\eta k_{2j}}, \quad k_{0j} \equiv \frac{k_{1j}}{k_{2j}}.$$

If $N_m > N_{m1} > 0$, then $N_m > N_{mj} > 0, j = 1, 2$. We now examine the conditions for $N_m > N_{m1} > 0, m = 1, 2$.

By (5.2.10), $k_{0j} = (R_0 + w_{1j})/(R_0 + w_{2j})$. By $w_{mj} = \beta z_{mj}\Lambda^{\alpha}$ and the definitions of k_{0j} and b_j, we have

$$k_{02} - k_{01} = \frac{(z_{12}z_{21} - z_{11}z_{22})\beta\Lambda^{\alpha} + (z_{21} + z_{12} - z_{11} - z_{22})R_0}{(R_0 + w_{21})(R_0 + w_{22})}\beta\Lambda^{\alpha}. \tag{5.2.15}$$

It can be seen that $z_{11}z_{22} < z_{12}z_{21}$ under $b_2 > 0$. If we assume $z_{21} + z_{12} > z_{11} + z_{22}$, then by (5.2.14) the requirements, $N_m > N_{m1} > 0$, are guaranteed if

$$\begin{gathered} k_{02}N_1 + N_2 > M_2 + M_1 > k_{01}N_1 + N_2, \\ k_{01}k_{02}N_1 + k_{02}N_2 > k_{01}M_2 + k_{02}M_1 > k_{01}k_{02}N_1 + k_{01}N_2. \end{gathered} \tag{5.2.16}$$

We see that it is difficult to explicitly interpret economic conditions for $N_m > N_{m1} > 0$. Summarizing the above discussion, we get the following proposition.

Proposition 5.2.1.
Let the two groups have the same propensities and let $A_{11} / A_{12} > A_{21} / A_{22}$ be held. In the case of $b_2 > 0$, (5.2.11) has a single solution, R_0; in the case of $b_2 < 0$, $b_1 > 0$ and $b_1^2 + b_2 > 0$, it has two positive solutions. For any positive solution, R_0, of (5.2.11), if $N_m > N_{mj} > 0$ are satisfied, then the dynamic system has a unique equilibrium corresponding to the given solution, R_0. We may determine the equilibrium values of the variables by the following procedure: Λ by (5.2.9) $\to$ r and w_{mj}, $m, j = 1, 2$, by (5.2.6) $\to$ R_0 by (5.2.11) $\to$ k_{mj} by (5.2.10) $\to$ R by (5.2.3) $\to$ R_j by (5.2.13) $\to$ N_{mj} by (5.2.14) $\to$ K by (5.2.8) $\to$ N_j^* by (5.1.1) $\to$ F_j by (5.1.3) $\to$ $K_j = \alpha F_j / r$ $\to$ y_{mj} by (5.1.5) $\to$ l_{mj}, c_{mj} and s_{mj} by (5.2.2) $\to$ U_{mj} by (5.1.6).

5.3 On Regional Dynamics with Multiple Groups

This chapter proposed an economic growth model with interregional trade and heterogeneous population. The model may be considered as a synthesis of the Solow-Swan one-sector growth model, the Kaldor-Pasinetti two-group model, and the model proposed in Chapter 2. The model examined the role of human capital, preferences, and regional locational conditions on the dynamic interdependence and economic growth among regions. This chapter is the first attempt to develop a compact synthetic economic framework with heterogeneous population in which the main ideas in growth theory and regional economics can be discussed. Irrespective of many obvious strict assumptions, the framework suggested in this chapter is, from a structural point of view, general in the sense that some well-known models in theoretical economics can be considered as special cases of our model. According to the literature in growth theory and regional economics, it is conceptually easy to refine, extend and generalize the model.

Appendix

A.5.1 Equilibria in the Case that the Propensities Are Different Between the Groups

The appendix provides the conditions for the existence of economic equilibria when the two groups have different propensities, i.e., $\eta_1 \neq \eta_2$, or $\xi_1 \neq \xi_2$, or $\lambda_1 \neq \lambda_2$. By (5.1.9), at equilibrium

$$\lambda_m \Omega_{mj} = k_{mj}, \quad m, j = 1, 2 \tag{5.A.1.1}$$

hold. Substituting (5.A.1.1) into (5.1.8) yields

$$l_{mj} R_j = \frac{\eta_m k_{mj}}{\lambda_m}, \quad c_{mj} = \frac{\xi_m k_{mj}}{\lambda_m}, \quad s_{mj} = \delta_k k_{mj}, \quad j, m = 1, 2. \tag{5.A.1.2}$$

Substituting (5.A.1.2) into (5.1.6) and then using (5.1.10), we get

$$k_m = \frac{k_{m1}}{k_{m2}} = \frac{A_{m2}}{A_{m1}} R^{\eta_m}, \quad m = 1, 2 \tag{5.A.1.3}$$

where $R \equiv R_1 / R_2$. By (5.A.1.3), we obtain

$$k_1 = A k_2^{\eta} \tag{5.A.1.4}$$

where

$$\eta \equiv \frac{\eta_1}{\eta_2}, \quad A \equiv \left(\frac{A_{21}}{A_{22}}\right)^{\eta} \frac{A_{12}}{A_{11}}.$$

Substituting c_{mj} and s_{mj} in (5.A.1.2) into (5.1.13) yields

$$F_1 + F_2 = \delta_1 K_{01} + \delta_2 K_{02} \tag{5.A.1.5}$$

where

$$K_{0m} \equiv k_{m1} N_{m1} + k_{m2} N_{m2}, \quad \delta_m \equiv \frac{\xi_m}{\lambda_m} + \delta_k .$$

Here, K_{0m} are the total capital stocks owned by group m. We have

$$K_{01} + K_{02} = K .$$

By (5.1.2) and $r = \alpha F_1 / K_1 = \alpha F_2 / K_2$ in (5.1.3), we have

$$\Lambda \equiv \frac{K_1}{N_1^*} = \frac{K_2}{N_2^*} .$$

By this equation and (5.1.3), we have

$$r = \frac{\alpha}{\Lambda^{\beta}}, \quad w_{mj} = \beta z_{mj} \Lambda^{\alpha}, \quad m, j = 1, 2 . \tag{5.A.1.6}$$

Substituting $\lambda_m l_{mj} R_j = \eta_m k_{mj}$ in (5.A.1.2) into the land constraint equations in (5.1.12) yields

$$\frac{\eta_1 k_{1j} N_{1j}}{\lambda_1} + \frac{\eta_2 k_{2j} N_{2j}}{\lambda_2} = L_j R_j, \quad j = 1, 2 . \tag{5.A.1.7}$$

Adding the two equations in (5.A.1.7), we get

$$R_0 = \frac{\eta_1 K_{01}}{\lambda_1 N} + \frac{\eta_2 K_{02}}{\lambda_2 N} \tag{5.A.1.8}$$

where we use (5.1.4).

By $r = \alpha F_j / K_j$ in (5.1.3) and $r = \alpha / \Lambda^\beta$ in (5.A.1.6), we have $F_j = K_j / \Lambda^\beta$. Substituting $F_j = K_j / \Lambda^\beta$ into (5.A.1.5) yields

$$\delta_1 K_{01} + \delta_2 K_{02} = \frac{K}{\Lambda^\beta} \tag{5.A.1.9}$$

where we use $K_1 + K_2 = K$. By $K_{01} + K_{02} = K$ and (5.A.1.9), we solve

$$K_{01} = \frac{1/\Lambda^\beta - \delta_2}{\delta_1 - \delta_2} K, \quad K_{02} = \frac{\delta_1 - 1/\Lambda^\beta}{\delta_1 - \delta_2} K. \tag{5.A.1.10}$$

By (5.A.1.1), (5.1.5) and the definitions of Ω_{mj}, we get

$$k_{mj}\left(\upsilon_m - r\right) = R_0 + w_{mj}, \quad m, j = 1, 2 \tag{5.A.1.11}$$

where $\upsilon_m \equiv 1/\lambda_m - \delta > 0$. By (5.A.1.6) and (5.A.1.11), we get

$$k_m = \frac{\Lambda^* + \beta z_{m1}}{\Lambda^* + \beta z_{m2}}, \quad m = 1, 2 \tag{5.A.1.12}$$

where $\Lambda^* \equiv R_0 / \Lambda^\alpha$. Substituting (5.A.1.12) into (5.A.1.4) yields

$$\Phi(\Lambda^*) \equiv \frac{\Lambda^* + \beta z_{11}}{\Lambda^* + \beta z_{12}} - \left(\frac{\Lambda^* + \beta z_{21}}{\Lambda^* + \beta z_{22}}\right)^\eta A = 0, \quad \infty > \Lambda^* > 0. \tag{5.A.1.13}$$

The function, $\Phi(\Lambda^*)$, has the following properties

$$\Phi(0) = \frac{z_{11}}{z_{12}} - \left(\frac{z_{21}}{z_{22}}\right)^\eta A, \quad \Phi(\infty) = 1 - A,$$

$$\frac{d\Phi}{d\Lambda^*} = \frac{z_{12} - z_{11}}{\left(\Lambda^* + \beta z_{12}\right)^2}\beta + \frac{\left(\Lambda^* + \beta z_{21}\right)^{\eta-1}}{\left(\Lambda^* + \beta z_{22}\right)^{\eta+1}}\left(z_{21} - z_{22}\right)\beta\eta A \,. \quad (5.A.1.14)$$

We see that the signs of $\Phi(0)$, $\Phi(\infty)$, and $d\Phi / d\Lambda^*$ are dependent on the parameter values. They may be either positive or negative. For instance, in the case of $z_{11} > (<) z_{12}$, $z_{21} < (>) z_{22}$ and $A > (<) 1$, we have $\Phi(0) > (<) 0$, $\Phi(\infty) < (>) 0$ and $d\Phi / d\Lambda^* < (>) 0$ for $\infty > \Lambda^* > 0$. We see that the equation, $\Phi(\Lambda^*) = 0$, in (5.2.13) has a unique positive solution. In the case of $z_{11} > (<) z_{12}$, $z_{21} < (>) z_{22}$ and $A < (>) 1$, we have $\Phi(0) > (<) 0$, $\Phi(\infty) > (<) 0$ and $d\Phi / d\Lambda^* < (>) 0$ for $\infty > \Lambda^* > 0$. We see that the equation, $\Phi(\Lambda^*) = 0$, in (5.A.1.13) has no solution in these cases. In the case of $z_{11} < z_{12}$, $z_{21} < z_{22}$, $z_{11} / z_{12} > A\left(z_{21} / z_{22}\right)^{\eta}$ and $A > 1$, we have that $\Phi(0) > 0$, $\Phi(\infty) < 0$ and the sign of $d\Phi / d\Lambda^*$ may be either positive or negative. In this case, the equation has at least one positive solution. The equation may have multiple solutions. It should be remarked that when the system has no solution, it does not mean that the problem has no meaningful solution. In this case, the economic activity may be concentrated in one region. Here, we will not examine each case in detail. In the remainder of this chapter, we are only concerned with the following special case.

Assumption 5.A.1.
$z_{11} > z_{12}$, $z_{21} < z_{22}$ and $A > 1$.

Under Assumption 5.A.1, the equation, (5.A.1.13), has a unique positive solution. The requirements of $z_{11} > z_{12}$ and $z_{21} < z_{22}$ imply that group 1's working efficiency is higher in region 1 than in region 2, and group 2's working efficiency is lower in region 1 than in region 2. The requirement of $A > 1$, i.e., $\left(A_{21} / A_{22}\right)^{\eta} > A_{11} / A_{12}$, is guaranteed, for instance, if the ratio of amenity levels that group 2 obtains in region 1 and region 2 is greater than unity, i.e., $A_{21} / A_{22} > 1$ and group 1 obtains the same level of amenity in the two regions, i.e., $A_{11} = A_{12}$.

Under Assumption 5.A.1, we get a unique positive value of Λ^* as a function of the parameters. By (5.A.1.12) and (5.A.1.3), we solve k_m and R as functions of the parameters. By (5.A.1.11) and (5.A.1.6), we get

$$k_{mj} = \frac{Z_{mj}\Lambda}{\upsilon_m \Lambda^{\beta} - \alpha}, \quad m,\ j = 1,\ 2 \tag{5.A.1.15}$$

where $Z_{mj} \equiv \Lambda^* + \beta z_{mj}$. By (5.A.1.8) and (5.A.1.10), we solve

$$K = \frac{(\delta_1 - \delta_2) N \Lambda \Lambda^*}{(1 - \delta_2 \Lambda^{\beta})\eta_1 / \lambda_1 - (1 - \delta_1 \Lambda^{\beta})\eta_2 / \lambda_2}. \tag{5.A.1.16}$$

By (5.1.4), $\Lambda^* = R_0 / \Lambda^{\alpha}$ and $R_1 = RR_2$, we get

$$R_1 = R\Lambda_0 \Lambda^*, \quad R_2 = \Lambda_0 \Lambda^* \tag{5.A.1.17}$$

where $\Lambda_0 \equiv \Lambda^* N / (RL_1 + L_2)$.

Substituting (5.A.1.17) and (5.A.1.15) into (5.A.1.7), we obtain

$$\frac{n_{11} N_{11} \Lambda^{\beta}}{\upsilon_1 \Lambda^{\beta} - \alpha} + \frac{n_{21} N_{21} \Lambda^{\beta}}{\upsilon_2 \Lambda^{\beta} - \alpha} = R, \quad \frac{n_{12} N_{12} \Lambda^{\beta}}{\upsilon_1 \Lambda^{\beta} - \alpha} + \frac{n_{22} N_{22} \Lambda^{\beta}}{\upsilon_2 \Lambda^{\beta} - \alpha} = 1 \tag{5.A.1.18}$$

where

$$n_{mj} \equiv \frac{\eta_m Z_{mj}}{\lambda_m \Lambda_0 L_j}, \quad m,\ j = 1,\ 2.$$

By $N_{m1} + N_{m2} = N_m$, $m = 1,\ 2$, and (5.A.1.18), we solve

$$N_{11} = \left\{ \frac{d_1}{\Lambda^{\beta}} - \frac{n_{12} n_{21} N_1}{\upsilon_1 \Lambda^{\beta} - \alpha} - \frac{n_{21} n_{22} N_2}{\upsilon_2 \Lambda^{\beta} - \alpha} \right\} (\upsilon_1 \Lambda^{\beta} - \alpha) d_0,$$

$$N_{12} = \left\{ -\frac{d_1}{\Lambda^{\beta}} + \frac{n_{11} n_{22} N_1}{\upsilon_1 \Lambda^{\beta} - \alpha} + \frac{n_{21} n_{22} N_2}{\upsilon_2 \Lambda^{\beta} - \alpha} \right\} (\upsilon_1 \Lambda^{\beta} - \alpha) d_0,$$

$$N_{22} = \left\{ \frac{d_2}{\Lambda^{\beta}} - \frac{n_{12} n_{21} N_2}{\upsilon_2 \Lambda^{\beta} - \alpha} - \frac{n_{11} n_{12} N_1}{\upsilon_1 \Lambda^{\beta} - \alpha} \right\} (\upsilon_2 \Lambda^{\beta} - \alpha) d_0$$

$$N_{22} = \left\{ \frac{d_2}{\Lambda^\beta} - \frac{n_{12}n_{21}N_2}{\upsilon_2\Lambda^\beta - \alpha} - \frac{n_{11}n_{12}N_1}{\upsilon_1\Lambda^\beta - \alpha} \right\} (\upsilon_2\Lambda^\beta - \alpha) d_0 \quad (5.A.1.19)$$

where

$$d_0 \equiv \frac{1}{n_{11}n_{22} - n_{12}n_{21}}, \quad d_1 \equiv n_{21} + n_{22}R, \quad d_2 \equiv n_{11} + n_{12}R.$$

Substituting (5.A.1.16), (5.A.1.15) and (5.A.1.19) into (5.1.11) yields

$$\Omega(\Lambda) \equiv \frac{D_0}{\Lambda^\beta} + \frac{D_1}{\upsilon_1\Lambda^\beta - \alpha} - \frac{D_2}{\upsilon_2\Lambda^\beta - \alpha} -$$

$$\frac{(\delta_1 - \delta_2)N\Lambda^*}{(\delta_1\Lambda^\beta - 1)\eta_2/\lambda_2 - (\delta_2\Lambda^\beta - 1)\eta_1/\lambda_1} \frac{1}{d_0} = 0 \quad (5.A.1.20)$$

where

$$D_0 \equiv d_1Z_{11} - d_1Z_{12} + d_2Z_{22} - d_2Z_{21},$$
$$D_1 \equiv -n_{12}n_{21}N_1Z_{11} + n_{11}n_{22}N_1Z_{12} + n_{11}n_{12}N_1Z_{21} - n_{11}n_{12}N_1Z_{22},$$
$$D_2 \equiv -n_{22}n_{21}N_2Z_{11} + n_{21}n_{22}N_2Z_{12} + n_{11}n_{12}N_2Z_{21} - n_{21}n_{12}N_2Z_{22} \quad (5.A.1.21)$$

For the variables to be meaningful, from (5.A.1.10) and (5.A.1.15) we see that in the case of $\delta_1 > \delta_2$ it is necessary to require

$$\min\left\{\frac{\upsilon_1}{\alpha}, \frac{\upsilon_2}{\alpha}, \delta_1\right\} > \frac{1}{\Lambda^\beta} > \delta_2. \quad (5.A.1.22)$$

As $\upsilon_1/\alpha > \delta_1$, the left hand side of (5.A.1.22) is given by

$$\min\left\{\frac{\upsilon_2}{\alpha}, \delta_1\right\} > \frac{1}{\Lambda^\beta}.$$

Summarizing the discussion in the appendix, we have the following proposition.

Proposition 5.A.1.

Let $\delta_1 > \delta_2$ and Assumption 5.2.1 be satisfied. For any solution of (5.2.20) which satisfies (5.A.1.22), the dynamic system has a single equilibrium. The number of economic equilibria is the same as that of meaningful solutions of (5.A.1.20). The equilibrium values of the variables are given by the following procedure: Λ^* by (5.A.1.13) $\rightarrow$ Λ by (5.A.1.20) $\rightarrow$ N_{mj}, m, $j = 1, 2$, by (5.A.1.19) $\rightarrow$ R_j by (5.A.1.17) $\rightarrow$ k_{mj} by (5.A.1.15) $\rightarrow$ K by (5.A.1.16) $\rightarrow$ $R_0 = \Lambda^* \Lambda^\alpha$ $\rightarrow$ r and w_{mj} by (5.A.1.6) $\rightarrow$ N_j^* by (5.1.1) $\rightarrow$ F_j by (5.1.3) $\rightarrow$ $K_j = \alpha F_j / r$ $\rightarrow$ y_{mj} by (5.1.5) $\rightarrow$ l_{mj}, c_{mj}, and s_{mj} by (5.A.1.2) $\rightarrow$ U_{mj} by (5.1.6).

It is difficult to explicitly interpret the conditions for existence of economic equilibria when the two groups have different propensities.

6 Urban and Rural Equilibrium Structures

The models so far proposed in this book neglect spatial structures of regional economies. Since space may appear in spatial models either in 'discrete' terms like in the regional models in the previous chapter or in 'continuous' terms like in the well-known Alonso model, it is important to build analytical frameworks for both cases.

Von Thünen developed a theory of the location of agricultural products in concentric rings around a central market. As pointed out by Blaug (1985), the history of location theory begins with the publication of von Thünen's *Der Isolierte Staat* (The Isolated State) in 1826. Von Thünen was concerned with the problem of agricultural activities which center on the competition of land use. He provided the classical analysis of allocation of land among competing agricultural activities. Von Thünen started to construct the isolated state economy as follows:

> Consider a very large town in the center of a fertile plain which does not contain any navigable rivers or canals. The soil of the plain is assumed to be of uniform fertility which allows cultivation everywhere. At a great distance the plain ends in an uncultivated wilderness, by which this state is absolutely cut off from the rest of the world.
> This plain is assumed to contain no other cities but the central town and in this all manufacturing products must be produced; the city depends entirely on the surrounding country for its supply of agricultural products.
> All mines and mineral deposits are assumed to be located right next to the central town.
> It was assumed that there was only one form of transportation, the horse and wagon, operated by the farmer at his own expense. There were no multiple freight rates depending on the commodities sipped.

The question now is: How under these circumstances will agriculture be developed and how will the distance from the city affect agricultural methods when these are chosen in the optimal manner? He showed that the town would be surrounded by agricultural rings: each ring cultivates a specific crop associated with the highest bid rent over the ring. The bid rent is the revenue minus the cost of labor and the cost of transportation. Differences in land use and agricultural production were the result of the types and quantities of agricultural products needed in the city, the technology employed in the production and transportation of such commodities, and the endeavor of each farmer to maximize his land rent by producing commodities. Various kinds of agricultural products are grown in concentric circles around the city, the exact location at which each product is raised being determined by the cost of transporting it to the city. For instance, in the area near to the city vegetables and fresh milk are produced and land cultivation is intensive because of the high price of land. Farther from the city, wheat is produced by the enclosure system. In the outermost circles the land is used for grazing.

Using the simple model, von Thünen examined many important issues in spatial analysis such as the effect of location on prices and land rent, the effect of urban demand fluctuations on the corresponding agricultural area, and interactions between the city and its hinterland. The model explains a number of important issues related to agricultural production locations in a simple manner. It explains the existence of agricultural specification even in the absence of climatic and fertility differences and the decrease of rent with distance from markets. It explains the relationship among distance of farms from the market, prices received by farmers for their products and land rent. The price which a farmer obtained for a given unit of commodity was equal to its price at the market minus the cost of shipping it to the market. The cost of transportation increased with distance from the market. The land rent was dependent on location relative to the market. The land rent nearest the city was the highest. The land rent decreased as location was further away from the city. It is a fertile model in the sense that it can be extended and modified in multiple ways. A key concept in land use theory is the concept of land rent. Ricardo and von Thünen wrote about agricultural land rent.

Isard (1956) noticed that the ideas of von Thünen could be reinterpreted in the context of urban land-use. Since then, there are some works, which mark a conceptual transition between the earlier work of land economists and modern urban economists (Beckmann, 1957, Mohring, 1961, Wingo, 1961, Muth, 1961, 1969). The availability of a well-developed microeconomic theory provided the theoretical foundations for the growth of modern urban economics. Beckmann (1957) studied the determination of equilibrium residential land rents and quantities in a monocentric city, where all employment and services are concentrated in the CBD (central business district) surrounded by a residential area. But it may be argued that the development of modern urban economics has been strongly influenced by the work of Alonso (1964). In his bid-rent theory, Alonso pioneered the adaptation of von Thünen's work on an urban context. The central market was replaced by a central business district, and agricultural products by alternative urban lands. According to Papageorgiou and Pines (1999), Alonso's greatest contribution was his proposal for the matching between spatial analysis and microeconomic theory that was necessary for the development of modern urban economics. Similar to the Solow-Swan model for neoclassical growth model, the Alonso model provides a simple mathematical structure based on which many articles have been published. The model has inspirited further development of urban economics mainly because of its logical consistency, simplicity, possible extensions and rich implications for important phenomena. Since the publication of Alonso's seminal work, urban economic has become an established field through the works of urban economists.

The Alonso model is concerned with urban land use and market land prices. In its simple form, the model of monocentric city assumes that all economic activities are concentrated in central business districts, which are surrounded by residential suburbs. It involves a density of consumers, identical with respect to income and tastes. Preferences are defined over the consumption of a composite good, which is found at the CBD and land. Land is considered as a commodity. Each consumer must occupy land at one and only one location. Since the disposable income of consumers varies with distance from the center due to differences in transportation costs, so does consumption. In equilibrium consumption and prices are such that everyone has the same utility level. The model explains pattern formation of the residential land use around the CBD. The price of land, the density of land use, and the equilibrium locations of the urban population are endogenously determined by the model. The key role of analysis in this approach is the concept of 'bid rent', which represents the maximum rent each participant in the market can pay at each location. As a result of the different possible uses of a location, there is a whole set of bid rents for any given location. The equilibrium rent is the maximum of the bid rents, which is geographically represented as the upper envelope of bid rent curves.

The purpose of this chapter is to devise an analytical framework within which the basic ideas of the two approaches mentioned above can be combined. We contrive an equilibrium model of economic geography with two cities and one farm. The model provides equilibrium conditions of a perfectly competitive spatial economy. The economic system consists of one agricultural and two industrial sectors. It is assumed that all the households have an identical preference and that any possible costs for professional changes and people movement among cities and countryside are omitted. The cities and the countryside are different in amenity and technology. This chapter examines how division of labor and difference in amenity among the urban and rural areas may affect spatial distribution of economic variables. The model in this chapter is an extension of the model on spatial division of labor and economic geography established by Zhang (1993b). The remainder of this chapter is organized as follows. Section 6.1 defines the two-city and one farm economic geographic model. Section 6.2 provides the conditions for existence of equilibria. Section 6.3 examines the effects of changes in city 1's amenity on the economic structure. Section 6.4 concludes the chapter. The appendix proves the proposition in Section 6.2.

6.1 The Economic Structure with Urban and Rural Areas

This chapter is concerned with an economic system consisting of two cities, indexed by 1 and 2, respectively, and one farm. Each city consists of two parts - the CBD and a residential area. The locations of the CBDs are pre-specified points and all industrial activities are concentrated at the CBDs. Similarly to Zhang (1993b), this chapter features a linear two-city system on a homogeneous plain whose width is of unity. For simplicity, it is assumed that the two CBDs are one-side edged as shown in

Fig.6.1.1. This implies that the workers of each city can travel only in one direction. As shown in Fig. 6.1.1, the agricultural sector is assumed to be located between the two urban areas.

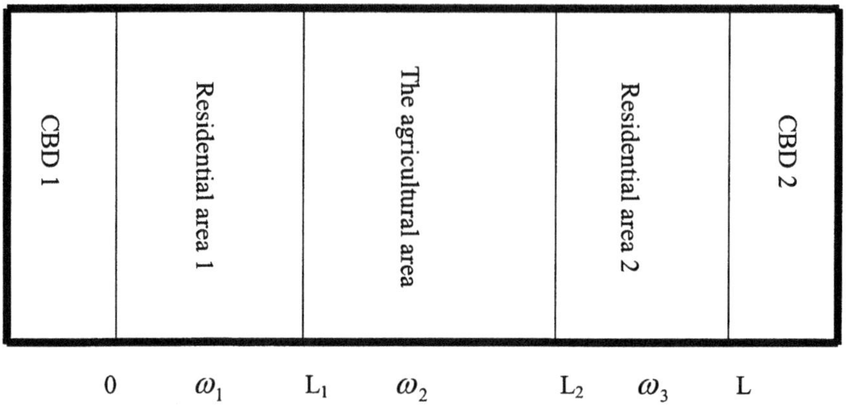

Fig. 6.1.1. The Two-City and One-Farm System

The system produces agricultural goods and two kinds of industrial commodities. The two commodities are indexed by 1 and 2, respectively. This chapter is concerned with an economic system with strict spatial division of labor. For simplicity, transportation costs of commodities are neglected. It is assumed that labor markets are characterized by perfect competition and that people are freely mobile among three professions. The labor force is homogeneous and agricultural good is selected to serve as numeraire, with all other prices being measured relative to its price. Wage rates may be different among different professions but identical within the same profession. The equilibrium of labor markets is guaranteed by the conditions that the households gets the same level of utility, irrespective of their location and profession. We define the following variables

L and N — the given labor force and territory size, respectively;

L_1 — the distance from CBD 1 to the boundary between residential area 1 and the agricultural area;

L_2 — the distance from CBD 1 to the boundary between the agricultural area and residential area 2;

ω_1, ω_2, and ω_3 — the dwelling location of city 1's residents, the farmers, and city 2's residents, $0 \le \omega_1 < L_1$, $L_1 \le \omega_2 < L_2$, and $L_2 \le \omega_2 < L$;

$R(\omega_j)$ — the land rent at location ω_j, $j = 1, 3$;

R_a — the land rent of the agricultural area;

N_j and N_a —city j's employment and the number of farmers, respectively;

F_j and F_a —the output of city j and the agricultural sector, respectively;

p_j —the price of commodity j; and

w_j and w_a —the wage rate in city j and wage rate of farmers, respectively.

Agricultural production

It is assumed that agricultural production is carried out by combination of labor force and land in the following way

$$F_a = L_a^{\alpha} N_a^{1-\alpha}, \quad 0 < \alpha < 1 \tag{6.1.1}$$

where L_a is land employed by the agricultural sector and α is a parameter. Profit maximization yields the following conditions

$$R_a = \frac{\alpha F_a}{L_a}, \quad w_a = \frac{(1-\alpha) F_a}{N_a}. \tag{6.1.2}$$

Let C_i and C_a denote total consumption of agricultural product by city j and farmers, respectively. The balance of demand for and supply of agricultural product is given by

$$C_1 + C_2 + C_a = F_a. \tag{6.1.3}$$

Industrial production

It is assumed that industrial commodities are produced with only one input, labor. Other possible inputs such as capital and land are neglected. Linear production functions of the two industrial sectors are specified as follows

$$F_j = z_j N_j, \quad j = 1, 2 \tag{6.1.4}$$

where z_j are city j's production efficiency parameters. The parameters, z_j, measures economic efficiency, and technology levels of city j in aggregated terms.

As there is only one input factor in each production sector, we have

$$w_j N_j = p_j F_j, \quad j = 1, 2$$

i.e.

$$w_j = p_j z_j, \quad j = 1, 2. \tag{6.1.5}$$

City j 's wage rate is equal to industry j 's product value per unit of labor input.

Let C_{ij} and C_{aj} denote total consumption of city j 's product by city i and by the farmers, respectively. The balances between demand for and supply of industrial commodities are given by

$$C_{11} + C_{21} + C_{a1} = F_1, \quad C_{12} + C_{22} + C_{a2} = F_2. \tag{6.1.6}$$

Behavior of consumers

It is assumed that the utility level of a household is dependent on four variables, consumption levels of the three commodities and of housing. Here, housing consumption is simply measured by the household's dwelling size. A typical household's utility function is specified as follows

$$U(\omega_j) = A_j c_1^{\xi_1}(\omega_j) c_2^{\xi_2}(\omega_j) c_a^{\xi_a}(\omega_j) c_h^{\xi_h}(\omega_j), \quad \xi_1, \xi_2, \xi_a, \xi_h > 0,$$
$$j = 1, 2, 3, \quad 0 \le \omega_1 < L_1, \quad L_1 \le \omega_2 < L_2, \quad L_2 \le \omega_3 < L \tag{6.1.7}$$

where $c_1(\omega_j)$, $c_2(\omega_j)$, $c_a(\omega_j)$ and $c_h(\omega_j)$ are, respectively, the consumption levels of commodity 1, commodity 2, agricultural good and housing of a household at location ω_j. The three parameters, A_1, A_2, and A_3 are respectively amenity levels of city 1, the countryside, and city 2 (e.g., Zhang, 1993b).

The consumer problem is defined by

$$\max U(\omega_j)$$

subject to

$$p_1 c_1(\omega_j) + p_2 c_2(\omega_j) + p_a c_a(\omega_j) + R(\omega_j) c_h(\omega_j) = y(\omega_j)$$
$$j = 1, 2, 3 \tag{6.1.8}$$

where

$$y(\omega_1) = w_1 - \tau\omega_1, \;\; 0 \le \omega_1 < L_1, \;\; y(\omega_2) = w_a, \;\; L_1 \le \omega_2 < L_2,$$
$$y(\omega_3) = w_2 - \tau\omega_3, \;\; L_2 \le \omega_3 < L$$

where τ is travel cost per unity of distance. Here, $\tau\omega_1$ and $\tau(L - \omega_3)$ are the total traveling cost per household between the dwelling site and the CBDs in urban areas 1 and 2, respectively. As wages are only income sources and savings are omitted, a household's wage is distributed among travel cost, industrial goods purchases and land rent.

The optimal problem has the following unique solution

$$c_1(\omega_j) = \frac{\xi\xi_1 y(\omega_j)}{p_1}, \;\; c_2(\omega_j) = \frac{\xi\xi_2 y(\omega_j)}{p_2}, \;\; c_a(\omega_j) = \frac{\xi\xi_a y(\omega_j)}{p_a},$$
$$c_h(\omega_j) = \frac{\xi\xi_h y(\omega_j)}{R(\omega_j)} \tag{6.1.9}$$

where

$$\xi \equiv \frac{1}{\xi_1 + \xi_2 + \xi_a + \xi_h}.$$

Let $n(\omega_j)$ denote the residential density at dwelling site ω_j. According to the definitions of n and c_h, we have

$$n(\omega_1) = \frac{1}{c_h(\omega_1)}, \;\; 0 \le \omega_1 < L_1, \;\; n(\omega_3) = \frac{1}{c_h(\omega_3)}, \;\; L_2 \le \omega_3 < L. \tag{6.1.10}$$

The population of each city is equal to the sum of its residents, i.e.

$$\int_0^{L_1} n(\omega_1)d\omega_1 = N_1, \;\; \int_{L_2}^{L} n(\omega_3)d\omega_3 = N_2. \tag{6.1.11}$$

The agricultural land which is equal to the territory size minus the total land used by the two cities is distributed between agricultural production and farmers' housing, i.e.

$$L_a + c_h(\omega_2)N_a = L_2 - L_1. \tag{6.1.12}$$

The conditions, (6.1.3) and (6.1.6), are rewritten as follows

$$\int_0^{L_1} n(\omega_1)c_j(\omega_1)d\omega_1 + \int_{L_2}^{L} n(\omega_3)c_j(\omega_1)d\omega_3 + (L_2 - L_1)c_j(\omega_2) = F_j,$$
$$j = 1, 2, a. \tag{6.1.13}$$

We completed constructing the model. The system consists of 22 variables, N_j, F_j, c_j, w_j, p_j, L_j $(j = 1, 2)$, F_a, N_a, w_a, R, c_h, c_a, n, U.

6.2 The Spatial Equilibrium Structure

This section proves that the 22 variables can be expressed as functions of z_1, z_2, L, N and the other parameters in the system. First, we note that the utility level of households is identical, irrespective of their dwelling location and professions. Using (6.1.7), (6.1.9), and

$$U(\omega_1) = U(0), \ \ 0 \le \omega_1 < L_1, \ \ U(\omega_3) = U(L), \ \ L_2 \le \omega_3 < L$$

we get

$$\frac{R(\omega_1)}{R_a} = \left\{\frac{y(\omega_1)}{y(L_1)}\right\}^{1/\xi\xi_h}, \ \ 0 \le \omega_1 < L_1,$$
$$\frac{R(\omega_3)}{R_a} = \left\{\frac{y(\omega_3)}{y(L_2)}\right\}^{1/\xi\xi_h}, \ \ L_2 \le \omega_3 < L \tag{6.2.1}$$

where

$$y(L_j) = w_j - \tau L_j, \ \ j = 1, 2.$$

Each urban area's land rent declines as it is further from its CBD. As the households from the same city earn the same wage rate, we see that this conclusion is quite reasonable.

Using (6.1.7), (6.1.9)

$$U(\omega_1) = U(\omega_2), \quad at\ \omega_1 = \omega_2 = L_1,$$
$$U(\omega_2) = U(\omega_3), \quad at\ \omega_2 = \omega_3 = L_2$$

We have

$$w_1 - \tau L_1 = \left(\frac{A_2}{A_1}\right)^{\xi} w_a, \quad w_2 - \tau(L - L_2) = \left(\frac{A_2}{A_3}\right)^{\xi} w_a. \tag{6.2.2}$$

If the level of amenity in the countryside is lower than in the levels in the cities, farmers' wage rate may be higher than workers in the cities in a competitive equilibrium. In reality one may find that urban workers may enjoy higher amenities and earn higher wage rates than farmers. This is, among many other factors, due to the fact that labor force has practically different human capital. In particular, educated people tend to be concentrated in cities. This suggests that in order to explain the difference in wages and amenities between urban workers and farmers it is necessary to further classify labor force into different classes according to education and human capital.

The following proposition shows that the system has a unique equilibrium under certain conditions.

Proposition 6.2.1.
If $A_3 > A_1$ and $\xi_1 < \xi_2$, the system has a unique equilibrium.

The above proposition is proved in the appendix.

The requirement, $A_3 > A_1$, implies that city 2's amenity level is higher than city 1. The condition of $\xi_1 < \xi_2$ states that the propensity to consume city 1's product is lower than city 2's product. It can be seen that the system also has a unique equilibrium if $A_3 < A_1$ and $\xi_1 > \xi_2$. It should be remarked that the uniqueness of equilibrium is not guaranteed in the case of $A_3 < A_1$ and $\xi_1 < \xi_2$ (or $A_3 > A_1$ and $\xi_1 > \xi_2$). In the remainder of this chapter, we require: $A_3 > A_1$ and $\xi_1 < \xi_2$.

Let us examine the case of $A_3 = A_1$, $\xi_1 = \xi_2$, and $\xi_a = 0$. The two cities have identical amenity, the propensities to consume the two industrial commodities are identical and there is no demand for the agricultural product. The equilibrium is given by

$$L_1 = L_2 = \frac{L}{2}, \quad N_1 = N_2 = \frac{N}{2}.$$

At the equilibrium, the agricultural sector disappears from the system, the boundary between the two urban areas is located at the middle point of the two cities, and the two cities employ the same number of labor.

6.3 City 1 's Amenity and the Economic Structure

We now examine the effects of changes in city 1 's amenity, A_1, on the equilibrium structure of the spatial economy. Taking derivatives of (6.A.1.13) with respect to A_1 yields

$$M'\frac{dW_2}{dA_1} = \frac{\xi\xi_2\alpha_1\tau\xi_h^2 L_1}{\xi_a\omega_2 A_1} \tag{6.3.1}$$

where we use (6.A.1.12) and (6.2.2), $W_2 = w_2 / w_a$, and $M' > 0$ is defined in (6.A.1.13). As city 1 's amenity is improved, the ratio of city 1 's and farmers' wage rates is increased. From (6.A.1.11) we obtain

$$\frac{dW_1}{dA_1} = -\frac{W_1}{\xi_h A_1} + \frac{(A_1 / A_2)^{1/\xi_h}(W_2 / W_1)^{1/\xi\xi_h - 1}\xi_1}{\xi\xi_2\xi_h}\frac{dW_2}{dA_1} \tag{6.3.2}$$

where $W_1 = w_1 / w_a$. In general, dW_1 / dA_1 may be either positive or negative. If dW_2 / dA_1 is small, then dW_1 / dA_1 tends to be negative. As an improvement in city 1 's amenity makes city 1 more attractive in comparison to the rural area and city 2, city 1 's wage rate tends to be reduced in the term of the wage rate in the agricultural sector.

Taking derivatives of (6.A.1.10) with respect to A_1 yields

$$\frac{(1-\alpha)W}{\alpha w_a}\frac{dw_a}{dA_1} = \left\{\frac{\xi_1}{W_1} + \frac{\xi_2}{W_2} + (1-\alpha)\xi_a\right\}\left(\frac{dW_1}{dA_1} - \frac{dW_2}{dA_1} + \frac{\xi A_2^{\xi}}{A_1^{\xi+1}}\right)$$
$$+ \left(\frac{\xi_1}{W_1^2}\frac{dW_1}{dA_1} + \frac{\xi_2}{W_2^2}\frac{dW_2}{dA_1}\right)(W_2 - W_1 + A) \tag{6.3.3}$$

where

$$W \equiv \left\{\frac{\xi_1}{W_1} + \frac{\xi_2}{W_2} + (1-\alpha)\xi_a\right\}(W_2 - W_1 + A).$$

From (6.3.2) and (6.2.2) we have

$$\frac{dW_1}{dA_1} - \frac{dW_2}{dA_1} + \frac{\xi A_2^{\xi}}{A_1^{\xi+1}} = -\frac{(1-\xi_h)\xi A_2^{\xi} / A_1^{\xi} - \tau L_1}{\xi_h A_1}$$
$$+ \left\{\frac{(A_1 / A_2)^{1/\xi_h} (W_2 / W_1)^{1/\xi\xi_h - 1}}{\xi\xi_a\xi_2}\xi_1 - 1\right\}\frac{dW_2}{dA_1}. \qquad (6.3.4)$$

The right-hand side of (6.3.4) is negative when dW_2 / dA_1 is small. Examining each term in the right-hand side in (6.3.3), one may conclude that when dW_2 / dA_1 is small, dw_a / dA_1 may be negative. Otherwise, dw_a / dA_1 may be either positive or negative. Summarizing the above discussion, one sees that it is only when dW_2 / dA_1 is small, it is possible to explicitly judge the signs of dW_1 / dA_1 and dw_a / dA_1.

In the remainder of this section, it is assumed that dW_2 / dA_1 is small so that $dW_1 / dA_1 < 0$ and $dw_a / dA_1 < 0$ are held. If this requirement is not satisfied, it is difficult to explicitly judge the effects of changes in A_1. From $W_1 = w_1 / w_a$, $W_2 = w_2 / w_a$, and (6.1.5), we have

$$\frac{dw_j}{dA_j} < 0, \quad \frac{dp_j}{dA_j} < 0, \quad j = 1, 2. \qquad (6.3.5)$$

An improvement in city 1's amenity reduces wage rates (in the term of agricultural product) in the three sectors and reduces the prices of the two industrial commodities.

Taking derivatives of (6.A.1.14) with respect to A_1 together with (6.2.2) and (6.3.2) yields

$$\frac{dL_1}{dA_1} = -\frac{\tau L_1 + (1 - \xi\xi_h)w_a A_2^{\xi} / A_1^{\xi}}{\tau\xi_h A_1} + \frac{L_1}{w_a}\frac{dw_a}{dA_1}$$
$$+\frac{\xi_1 w_a (A_1 / A_2)^{1/\xi_h} (W_2 / W_1)^{1/\xi\xi_h - 1}}{\tau\xi\xi_h\xi_2}\frac{dW_2}{dA_1},$$
$$\frac{dL_2}{dA_1} = \frac{w_a}{\tau}\frac{dW_2}{dA_1} + \frac{L_2}{w_a}\frac{dw_a}{dA_1}. \qquad (6.3.6)$$

City 1's urban area is decreased but city 2's urban area, $L - L_2$, is expanded as city 1's amenity is increased. From (6.A.1.1) and (6.3.6) we have

$$\frac{1}{R_a}\frac{dR_a}{dA_1} = -\frac{(1-\alpha)L_2}{\alpha w_a}\frac{dw_a}{dA_1}, \quad \frac{1}{\alpha\alpha_1}\frac{dL_a}{dA_1} = \frac{dL_2}{dA_1} - \frac{dL_1}{dA_1},$$
$$\frac{1}{N_a}\frac{dN_a}{dA_1} = \frac{1}{(1-\alpha)R_a}\frac{dR_a}{dA_1} + \frac{1}{L_a}\frac{dL_a}{dA_1},$$
$$\frac{1}{c_h(\omega_2)}\frac{dc_h(\omega_2)}{dA_1} = -\frac{1}{(1-\alpha)R_a}\frac{dR_a}{dA_1} < 0. \qquad (6.3.7)$$

The land rent in the agricultural area is increased. The land and labor force employed by the agricultural sector may be either increased or decreased. The last equation in (6.3.7) implies that an increase in city 1's amenity reduces dwelling size of the farmers.

From (6.A.1.3) we have

$$\frac{dN_1}{dA_1} = \frac{N_1}{R_a}\frac{dR_a}{dA_1} - \left[\frac{L_1}{w_1} - \frac{dL_1}{dA_1}\right]\frac{R_a w_1}{\xi\xi_h^3 y(L_1)^{1/\xi\xi_h - 1}}\frac{dw_1}{dA_1},$$
$$\frac{dN_2}{dA_1} = \frac{N_2}{R_a}\frac{dR_a}{dA_1} - \left[\frac{L_2}{w_2} - \frac{dL_2}{dA_1}\right]\frac{R_a w_2}{\xi\xi_h^3 y(L_2)^{1/\xi\xi_h - 1}}\frac{dw_2}{dA_1}. \qquad (6.3.8)$$

From (6.1.1), (6.1.4) and the above results, we can calculate the effects of changes in A_1 on dF_a / dA_1, dF_1 / dA_1, and dF_2 / dA_1. From (6.2.1) we calculate $R(\omega_1)$ for $0 \le \omega_1 < L_1$ and $R(\omega_3)$ for $L_2 \le \omega_1 < L$. From (6.1.9) and (6.1.10) we may obtain the effects on the consumption components of all the households and the residential densities in the two urban areas.

6.4 On Spatial Economic Structure

This chapter proposed an economic geographic model of the two cities and one farm. Each city produces only one commodity and the farm supplies agricultural goods. Amenities are spatially different among the cities and farming area. It is shown that the system has a unique equilibrium under certain conditions. The chapter also examined the impact of changes in city 1's amenity on the spatial economic structure.

Appendix

A.6.1 Proving Proposition 6.2.1

The appendix proves Proposition 6.2.1. First, from (6.1.12), and

$$c_h(\omega_2) = \frac{\xi\xi_h w_a}{R_a}, \quad R_a = \frac{\alpha F_a}{L_a}, \quad w_a = \frac{(1-\alpha)F_a}{N_a}$$

we get

$$L_a = \alpha\alpha_1(L_2 - L_1), \quad N_a = L_a\left(\frac{R_a}{\alpha}\right)^{1/(1-\alpha)}, \quad w_a = \sigma_0 R_a^{-\alpha/(1-\alpha)},$$

$$c_h(\omega_2) = \frac{\xi\xi_h L_a}{\alpha N_a}(1-\alpha), \quad L_1 < \omega_2 \le L_2 \tag{6.A.1.1}$$

where

$$\alpha_0 \equiv (1-\alpha)\alpha^{\alpha/(1-\alpha)}, \quad \alpha_1 \equiv \frac{1}{\alpha + (1-\alpha)\xi\xi_h}.$$

We see that the variables, L_a, N_a, $c_h(\omega_2)$ and w_a, of the agricultural sector can be expressed as functions of the variables, L_1, L_2 and R_a. From (6.1.9), (6.1.10) and (6.2.1), we have

$$n(\omega_1) = \frac{\{y(\omega_1)/y(L_1)\}^{1/\xi\xi_h} R_a}{\xi\xi_h y(\omega_1)}, \quad 0 \le \omega_1 < L_1,$$

$$n(\omega_3) = \frac{\{y(\omega_3)/y(L_2)\}^{1/\xi\xi_h} R_a}{\xi\xi_h y(\omega_3)}, \quad L_2 \le \omega_3 < L. \tag{6.A.1.2}$$

Substituting (6.A.1.2) into (6.1.11) yields

$$N_j = \frac{\{w_j / y(L_j)\}^{1/\xi\xi_h} - 1}{\tau\xi_h^2} R_a. \tag{6.A.1.3}$$

Substituting (6.1.9) and (6.A.1.2) into (6.1.13) together with (6.1.4) and (6.1.5) yields

$$G_1 + G_2 + (L_2 - L_1)\xi w_a = \frac{z_j p_j N_j}{\xi_j}, \quad j = 1, 2,$$

$$G_1 + G_2 + (L_2 - L_1)\xi w_a = \frac{F_a}{\xi_a} \tag{6.A.1.4}$$

where

$$G_j \equiv \left\{\frac{\xi R_a}{(\xi + \xi_h)\tau\xi_h y(L_j)^{1/\xi\xi_h}}\right\}\left\{w_j^{1+1/\xi\xi_h} - y(L_j)^{1+1/\xi\xi_h}\right\}. \tag{6.A.1.5}$$

From (6.A.1.4) we have, $w_1 N_1 / \xi_1 = w_2 N_2 / \xi_2 = F_a / \xi_a$, or

$$N_j = \frac{\xi_j F_a}{\xi_a w_1}, \quad j = 1, 2. \tag{6.A.1.6}$$

From (6.A.1.1), $F_a = \alpha_1 R_a (L_2 - L_1)$. With this equation, (6.A.1.6) and (6.A.1.3) we obtain

$$\left\{\frac{w_j}{y(L_j)}\right\}^{1/\xi\xi_h} = \frac{\tau\alpha_1\xi_j\xi_h^2}{\xi_a w_j}(L_2 - L_1) + 1. \tag{6.A.1.7}$$

Substituting $N_a = (1 - \alpha)F_a / w_a$ and (6.A.1.6) into $N_1 + N_2 + N_a = N$ yields

$$\left\{\frac{\xi_1}{\xi_a w_1} + \frac{\xi_2}{\xi_a w_2} + \frac{1-\alpha}{w_a}\right\}(L_2 - L_1)\alpha_1 R_a = N. \tag{6.A.1.8}$$

Substituting

$$y(L_1) = \left(\frac{A_2}{A_1}\right)^{\xi} w_a, \quad y(L_2) = \left(\frac{A_2}{A_3}\right)^{\xi} w_a, \quad w_a = \sigma_0 R_a^{-\alpha/(1-\alpha)}$$

into (6.A.1.7) and (6.A.1.8) yields

$$\left(\frac{A_1}{A_2}\right)^{1/\xi_h} W_1^{1/\xi\xi_h} = \left(\frac{W_2}{W_1} - 1 + \frac{A}{W_1}\right)\frac{\alpha_1\xi_1\xi_h^2}{\xi_a} + 1,$$

$$\left(\frac{A_3}{A_2}\right)^{1/\xi_h} W_2^{1/\xi\xi_h} = \left(-\frac{W_2}{W_1} + 1 + \frac{A}{W_2}\right)\frac{\alpha_1\xi_2\xi_h^2}{\xi_a} + 1, \tag{6.A.1.9}$$

$$w_a^{(1-\alpha)/\alpha} = \frac{\tau N \alpha_0^{(1-\alpha)/\alpha}}{(W_2 - W_1 + A)\{\xi_1/\xi_a W_1 + \xi_2/\xi_a W_2 + (1-\alpha)\}\alpha_1} \tag{6.A.1.10}$$

where

$$W_j \equiv \frac{w_j}{w_a}, \quad j = 1, 2, \quad A \equiv \left(\frac{A_2}{A_1}\right)^{\xi} - \left(\frac{A_2}{A_3}\right)^{\xi}.$$

The two equations in (6.A.1.9) have only two independent variables W_1 and W_2. From (6.A.1.9) we solve

$$\frac{(A_1/A_2)^{1/\xi_h} W_1^{1/\xi\xi_h} - 1}{(A_3/A_2)^{1/\xi_h} W_2^{1/\xi\xi_h} - 1} = \frac{\xi_1}{\xi_2}. \tag{6.A.1.11}$$

In the case of $\xi_1 = \xi_2$, $w_1 = (A_3/A_1)^{\xi} w_2$. This states that if the propensities to consume the two industrial commodities are equal, then the sign of the difference in wage rates between the two cities is the opposite to that in amenity levels between the two cities. The city with the higher amenity level has a lower wage rate than the other one. Moreover, in the case of $\xi_1 = \xi_2$ and $A_1 = A_3$, the wage rates in the two cities are identical, i.e., $w_1 = w_2$.

From (6.A.1.11) we obtain

$$\frac{W_1}{W_2} = f(W_2) \equiv \left(\frac{A_2}{A_1}\right)^{\xi} \left\{ \frac{\xi_1}{\xi_2}\left(\frac{A_3}{A_2}\right)^{1/\xi_h} + \frac{1 - \xi_1/\xi_2}{W_2^{1/\xi\xi_h}} \right\}^{\xi\xi_h} . \tag{6.A.1.12}$$

Substituting (6.A.1.12) into the second equation in (6.A.1.9) yields

$$M(W_2) \equiv \left(\frac{A_3}{A_2}\right)^{1/\xi_h} W_2^{1/\xi\xi_h} - \left\{1 - f(W_2) + \frac{A}{W_2}\right\} \frac{\alpha_1 \xi \xi_h^2}{\xi_a} = 0 . \tag{6.A.1.13}$$

Whether $M(W_2) = 0$ has a unique positive solution is dependent on the signs of A and $1 - \xi_1/\xi_2$. It is remarked that $A_3 > A_1$ implies $A > 0$. Under the requirement of $A > 0$ and $\xi_1 < \xi_2$, we have: $M(0) < 0, M(\infty) > 0$, and $M'(W_2) > 0$ for $0 < W_2 < +\infty$. Hence, $M(W_2) = 0$ has a unique positive solution. We can express W_2 as a function of the parameters in the system. It is now shown how all other variables in the system can be solved as functions of the system parameters.

The two variables, W_1 and w_a, are determined by $W_1 = W_2 f(W_2)$ and (6.2.12), respectively. The wage rates, w_1 and w_2, in the two cities are given by $w_1 = w_a W_1$ and $w_2 = w_a W_2$. We determined the wage rates, w_a, w_1 and w_2. The two boundaries, L_1 and L_2, are given by

$$L_1 = \frac{w_1 - (A_2/A_1)^{\xi} w_a}{\tau}, \quad L_2 = \frac{w_2 - (A_2/A_3)^{\xi} w_a}{\tau} . \tag{6.A.1.14}$$

The land rent of the agricultural area is given by $R_a = (\sigma_0 / w_a)^{1/\alpha - 1}$. The employment, N_1 and N_2, in the two cities are directly given by (6.A.1.3). The residential density, $n(\omega_1)$ and $n(\omega_2)$, in the two urban areas are determined by

(6.2.4). The land and labor force, L_a and N_a, employed the by agricultural sector are given by

$$L_a = \alpha\alpha_1(L_2 - L_1),\ N_a = \left(\frac{R_a}{\alpha}\right)^{1/(1-\alpha)} L_a.$$

The land rent distribution, $R(\omega_1)$ and $R(\omega_2)$, in the two urban areas are given by (6.2.1). The output, F_a, F_1, and F_2, of agricultural and industrial production are given by (6.1.1) and (6.1.4), respectively. The prices, p_1 and p_2, of the two industrial commodities are solved as $p_1 = w_1 / z_1$ and $p_2 = w_2 / z_2$. The consumption components of the households in the agricultural area and the two urban areas are obtained by (6.1.9).

Summarizing the above discussion, we proved Proposition 6.2.1.

7 Regional Economic Equilibrium with Two Groups

Regional location, interaction, economic structure and dynamic processes of various social and economic activities under different institutions are the main concerns of regional science. Regional economic analysis is to analyze regional location, regional demand and supply, prices of goods, services, distribution of immobile factors such as land and amenities among various activities, and interregional (monetary, material and information) flows (e.g., Isard, 1960). It is greatly desirable to develop a consistent framework to take all these important aspects of regional economies. From an economic structural point of view, it may be argued that two modelling frameworks are important in dealing with multiregional economic issues. The first is the interregional input-output analysis (e.g., Leontief, 1936, 1941, Isard, 1953). Although the input-output systems have proved effective for analyzing economic structure with complicated linkages among various sectors in multi-regional or multi-national systems, it may be argued that it is often analytically difficult to introduce endogenous behavior of households in an effective way. The second is equilibrium approach. There are many efforts on applying and extending the general equilibrium model to multiple regional systems. But as well argued by Scotchmer and Thisse (1992), equilibrium economics has failed in providing significant insights into the complexity of economic geography (see, e.g., Alonso, 1964, Beckmann, 1968, Greenhut, Norman and Hung, 1987, Krugman, 1991, Rauch, 1991, Fujita, 1989). We refer the comprehensive review, by Scotchmer and Thisse (1992), on the reasons that the current models do not seem to integrate the spatial factor satisfactorily.

It has now become clear that there are a number of potentially important spatial influences, such as public goods, amenities, different externalities, transportation costs, that may challenge the validity of competitive equilibrium theory for explaining a regionally heterogeneous economy. For instance, one difficult factor is the so-called capitalization, which implies that the price of land is interdependent with local amenities, economic agents' densities, transportation costs and other local variables or parameters. Although the significance of capitalization has been noticed by location theorists, it may be said that we still have no compact framework within which we can satisfactorily explain the issue. Indeed, the seminal paper on compensating regional variation in wages and rents by Roback (1982) has caused a wide interest among regional and urban economists to theoretically investigate how the value of location attributes is capitalized into wages and services. Since the publication of Roback's work, many empirical and theoretical studies have also shown that between urban areas wages may capitalize differences in amenity levels or living costs (e.g., Sitivanidou and Wheaton, 1992, Simon and Love, 1990, Bell, 1991, Voith, 1991). The purpose of this chapter is to discuss the issue within a perfectly competitive equilibrium framework.

The main contribution of this chapter is to generate a two-region economy within an equilibrium framework with constant return-to-scale economies and regional immobile factors of land and amenity. This chapter is organized as follows. Section 7.1 defines the two-region equilibrium model. Section 7.2 provides equilibrium conditions. Section 7.3 examines the effects of changes in the population of the farmer class on the economic structure. Section 7.4 analyzes the impact of changes in region 1's amenity for the landlord class on the economic structure. Section 7.5 concludes the chapter. This chapter is based on Zhang (1995c, 1995d).

7.1 The Agricultural Economy with Two Groups

We are concerned with an economic system consisting of two regions, indexed by $j = 1, 2$, respectively. Each region has only two resources, labor, and land. For simplicity of analysis, it is assumed that the given land quality and climates are homogenous within each region, but they are different between the regions.

The population is classified into two, farmer and landlord, classes. The households of the same class are identical in preference and human capital, but people of different classes may have different preferences and human capital. It is assumed that the landlord class does not work in field and lives on its land revenue. Farmers are paid in wages, being their only income resource. It is conceptually easy to relax this classification, allowing the existence of another class which owns a certain amount of land and work in its own land and some of the landlord class' land. For analytical convenience, we limit our discussion to the case of only two classes with strict division of labor and division of consumption.

Production of each region is classified into two, services and region-specified, sectors. The services sector produces regional services, such as tea houses and restaurants, housing, transportation, churches and schools, which can be consumed only by the producer region. Here, we emphasize the geographical character of services in our modeling. For convenience of analysis, we neglect possibility of tourist consumption. We omit any possibility that the region-specified goods are produced in the other region (due to technology, climate, land quality or any other reasons). The region-specified goods are consumed by the two regions. Perfect competition is assumed to prevail in markets both within each region and between the two regions, and commodities are traded without any barriers such as transport costs or tariffs. Labor and land markets are assumed to be perfectly competitive. Region 1's region-specified commodity is selected to serve as numeraire.

We introduce

$1,\ 2$ — subscript indexes for regions 1 and 2, respectively;
$s,\ n$ — subscript indexes for the service and region-specified sector, respectively;
L_j — region j' fixed land, $j = 1,\ 2$;
N_0 and N — the fixed population of landlord and farmer classes, respectively;
N_{0j} and N_j — the population of landlord and farmer classes in region j, respectively;
N_{jk} — the number of farmers employed by sector k in region j, $k = s,\ n$;
L_{jk} — the land employed by sector k in region j, $k = s,\ n$;
C_{0js} and C_{js} — the consumption levels of services by region j's landlord and farmer classes, respectively;
C_{0jq} and C_{jq} — the consumption levels of region j's region-specified goods by region q's landlord and farmer classes, respectively, $q = 1,\ 2$;
F_{jk} — the output of sector k in region j, $k = s,\ n$;
w_j — region j's wage rate;
p_j — the interregional identical price of region j's region-specified goods, $p_1 = 1$;
p_{js} — region j's service prices; and
R_j — region j's land rent.

The production functions of the national economy are specified as follows

$$F_{jk} = \left(q_{jk} L_{jk}\right)^{\alpha_{jk}} \left(z_{jk} N_{jk}\right)^{\beta_{jk}}, \quad \alpha_{jk} + \beta_{jk} = 1, \quad \alpha_{jk},\ \beta_{jk} > 0$$
$$j = 1,\ 2, \quad k = s,\ n$$

where q_{jk} and z_{jk} are, respectively, region j's land quality and human capital for sector k. We may consider z_{jk} dependent on working time and efficiency.

As q_{jk} and z_{jk} are fixed in this chapter, we rewrite the above production functions as follows

$$F_{jk} = \theta_{jk} L_{jk}^{\alpha_{jk}} N_{jk}^{\beta_{jk}}, \quad \theta_{jk} \equiv q_{jk}^{\alpha_{jk}} z_{jk}^{\beta_{jk}}. \tag{7.1.1}$$

We call the parameters θ_{jk} production efficiency of sector k in region j. We assume that the services sectors in the two regions have the same production function, i.e., $\alpha \equiv \alpha_{1s} = \alpha_{2s}$ and $\beta \equiv \beta_{1s} = \beta_{2s}$. The marginal conditions are given by

$$R_j = \frac{\alpha_{jn} p_j F_{jn}}{L_{jn}} = \frac{\alpha p_{js} F_{js}}{L_{js}}, \quad w_j = \frac{\beta_{jn} p_j F_{jn}}{N_{jn}} = \frac{\beta p_{js} F_{js}}{N_{js}}. \tag{7.1.2}$$

As the farmer class receives wages and owns no land, the net incomes, Y_j, of the two regions' farmers are given by

$$Y_j = w_j N_j. \tag{7.1.3}$$

We now define net incomes, Y_{0j}, of the two regions' landlords. Land properties may be distributed in multiple ways under various institutions. In this chapter, the land is assumed to be owned by the landlord class. To simplify the discussion, we specify the land revenue as follows. We assume that each landlord owns L_1 / N_0 amount of land in region 1 and L_2 / N_0 in region 2 and it is impossible to sell land but it is free to rent one's own land to farmers. The land revenue of landlord is given

$$R_0 = \frac{L_1 R_1 + L_2 R_2}{N_0}. \tag{7.1.4}$$

Each member of the landlord class receives R_0, irrespective of whether its dwelling site is located in region 1 or 2. Under the above specified land ownership distribution, the net incomes of the two regions' landlord class are given by

$$Y_{0j} = R_0 N_{0j}. \tag{7.1.5}$$

It is assumed that households' utility levels are dependent on the consumption levels of services and two country-specified goods. The utility functions are specified as follows

$$U_{0j} = A_{0j} \left(\frac{C_{0js}}{N_{0j}} \right)^{\eta_0} \left(\frac{C_{0j1}}{N_{0j}} \right)^{\xi_0} \left(\frac{C_{0j2}}{N_{0j}} \right)^{\lambda_0}, \quad \eta_0, \xi_0, \lambda_0 > 0,$$

$$U_j = A_j \left(\frac{C_{js}}{N_j} \right)^{\eta} \left(\frac{C_{j1}}{N_j} \right)^{\xi} \left(\frac{C_{j2}}{N_j} \right)^{\lambda}, \quad \eta, \xi, \lambda > 0 \tag{7.1.6}$$

where U_{0j} and U_j stand for the utility levels of region j's landlord and farmer classes, respectively. In (7.1.6), A_{0j} and A_j are called amenity levels of region j's landlord and farmer classes, respectively. They are considered as locational parameters or slow changing variables, such as infrastructures, regional cultures and climates, which may affect attractiveness of a location. As discussed in Chapter 2, this assumption can be relaxed by treating A_j as endogenous variables.

Each class makes decision on three variables. The financial budget constraints of the two classes are given by

$$C_{0j1} + p_2 C_{0j2} + p_{js} C_{0js} = Y_{0j}, \quad C_{j1} + p_2 C_{j2} + p_{js} C_{js} = Y_j. \tag{7.1.7}$$

Each household maximizes the utility subject to the budget constraint. The optimal problems have the following unique solutions

$$p_{js} C_{0js} = \rho_0 \eta_0 Y_{0j}, \quad C_{0j1} = \rho_0 \xi_0 Y_{0j}, \quad p_{js} C_{0j2} = \rho_0 \lambda_0 Y_{0j},$$
$$p_{js} C_{js} = \rho \eta Y_j, \quad C_{j1} = \rho \xi Y_j, \quad p_{js} C_{j2} = \rho \lambda Y_j \tag{7.1.8}$$

where

$$\rho_0 \equiv \frac{1}{\eta_0 + \xi_0 + \lambda_0}, \quad \rho \equiv \frac{1}{\eta + \xi + \lambda}. \tag{7.1.9}$$

As people are freely mobile between regions, the utility level of households from the same class should be equal, irrespective of which region they live, i.e.

$$U_{01} = U_{02}, \quad U_1 = U_2. \tag{7.1.10}$$

As services are consumed only by the producer region, we have

$$C_{0js} + C_{js} = F_{js}. \tag{7.1.11}$$

According to the definition of region-specified goods, the following demand-supply conditions are satisfied

$$C_{01j} + C_{1j} + C_{02j} + C_{2j} = F_{jn}. \tag{7.1.12}$$

It is assumed that land and the farmer class are fully employed in both regions, i.e.

$$L_{jn} + L_{js} = L_j, \quad N_{jn} + N_{js} = N_j. \tag{7.1.13}$$

By definitions, we have

$$N_{01} + N_{02} = N_0, \quad N_1 + N_2 = N. \tag{7.1.14}$$

We have thus built the model. The system has 44 independent variables, L_{jn}, L_{js}, N_{jn}, N_{js}, F_{jn}, F_{js}, C_{0js}, C_{0j1}, C_{0j2}, Y_{0j}, U_{0j}, C_{js}, C_{j1}, C_{j2}, Y_j, U_j, N_{0j}, N_j, p_{js}, w_j, R_j $(j = 1, 2)$, p_2, and R_0. The system has 49 equations.

The four equations in (7.1.7) are dependent on (7.1.8). Adding all the equations in (7.1.3) and (7.1.4), we have

$$Y_{01} + Y_{02} + Y_1 + Y_2 = F_{1n} + F_{2n} + F_{1s} + F_{2s}$$

where (7.1.4), (7.1.13), (7.1.14) and (7.1.2) are used. By (7.1.6) and (7.1.10), we have

$$\sum_j p_j \left(C_{01j} + C_{1j} + C_{02j} + C_{2j}\right) = F_{1n} + p_2 F_{2n}.$$

The above equation implies that one of the two equations in (7.1.12) is dependent on the other equations in the system. We see that the system also contains 44 independent equations. We now solve the equilibrium problem.

7.2 The Regional Economic Equilibrium Structure

This section examines under what conditions the economic system has equilibria. First, by (7.1.2) we have

$$\frac{L_{js}}{L_{jn}} = \frac{\alpha \beta_{jn}}{\beta \alpha_{jn}} \frac{N_{js}}{N_{jn}}, \quad j = 1, 2. \tag{7.2.1}$$

Substituting (7.1.7) into (7.1.6) and then using (7.1.9), we obtain

$$\frac{p_{1s}}{p_{2s}} = A_0, \quad \frac{w_1}{w_2} = A \tag{7.2.2}$$

in which (7.1.3) and (7.1.5) are used and

$$A_0 \equiv \left(\frac{A_{01}}{A_{02}}\right)^{1/\eta_0}, \quad A \equiv \left(\frac{A_2}{A_1}\right)^{\rho}\left(\frac{A_{01}}{A_{02}}\right)^{\eta\rho/\eta_0}. \tag{7.2.3}$$

The ratios of the two regions' service prices and wage rates are constant. The ratio, p_{1s} / p_{2s}, of region 1's and 2's services prices is positively related to the ratio, A_{01} / A_{02}, of the two regions' levels of the landlord class's amenities. The ratio, w_1 / w_2, of region 1's and 2's wage rates is positively related to the ratio, A_{01} / A_{02}, of the two regions' levels of the landlord class's amenities and negatively related to the ratio, A_1 / A_2, of the two regions' levels of the farmer class's amenities.

By $w_j = \beta p_{js} F_j / N_{js}$ in (7.1.2), (7.1.1), and (7.2.2), we have

$$\frac{L_{1s}}{N_{1s}} = \left(\frac{A}{A_0}\right)^{1/\alpha} \frac{L_{2s}}{N_{2s}}. \tag{7.2.4}$$

By $R_j = \alpha p_{js} F_{js} / L_{js}$, (7.1.1), (7.2.2) and (7.2.4), we get

$$\frac{R_1}{R_2} = \frac{A_0^{1/\alpha}}{A^{\beta/\alpha}}. \tag{7.2.5}$$

Substituting $p_{js} C_{0js} = \rho_0 \eta_0 Y_{0j}$ and $p_{js} C_{js} = \rho \eta Y_{0j}$ in (7.1.8) into $C_{0js} + C_{js} = F_{js}$ in (7.1.11) yields

$$\rho_0 \eta_0 Y_{0j} + \rho \eta Y_j = p_{js} F_{js}. \tag{7.2.6}$$

From (7.1.3)-(7.1.5), (7.2.6) and (7.1.2), we get

$$Y_j = \frac{\beta p_{js} F_{js} N_j}{N_{js}}, \quad Y_{0j} = \frac{d_j p_{js} F_{js} N_{01}}{\rho_0 \eta_0 L_{js}}, \quad j = 1, 2 \tag{7.2.7}$$

where

$$d_1 \equiv \left(L_1 + \frac{L_2 A^{\beta/\alpha}}{A_0^{1/\alpha}} \right) \frac{\alpha \rho_0 \eta_0}{N_0}, \quad d_2 \equiv \frac{d_1 A_0^{1/\alpha}}{A^{\beta/\alpha}}. \tag{7.2.8}$$

Substituting (7.2.7) into (7.2.6) yields

$$\frac{d_j N_{0j}}{L_{js}} + \frac{\rho \eta \beta N_j}{N_{js}} = 1. \tag{7.2.9}$$

Substituting $L_{jn} = L_j - L_{js}$ and $N_{jn} = N_j - N_{js}$ in (7.1.13) into (7.2.1) yields

$$\frac{N_j}{N_{js}} = \frac{\alpha \beta_{jn} L_j}{\alpha_{jn} \beta L_{js}} + \frac{\alpha_{jn} - \alpha}{\alpha_{jn} \beta}. \tag{7.2.10}$$

From (7.2.10) and (7.2.9), we solve N_{js} and L_{js} as functions of N_j and N_{01} as follows

$$L_{js} = \frac{d_j N_{01} + \alpha_0 L_j}{\alpha_0 + \beta_0}, \quad N_{js} = \frac{d_j N_{01} + \alpha_0 L_j}{\alpha \beta_{jn} L_j + d_j N_{01} (\alpha_{jn} - \alpha)} \alpha_{jn} \beta N_j \tag{7.2.11}$$

where

$$\alpha_0 \equiv \frac{\alpha \rho \eta \beta_{jn}}{\alpha_{jn}}, \quad \beta_0 \equiv 1 - \rho \eta \beta > 0.$$

By (7.1.13) and (7.2.11), we obtain

$$L_{jn} = \frac{\beta_0 L_j - d_j N_{01}}{\alpha_0 + \beta_0},$$

$$N_{jn} = \frac{\beta_0 L_j - d_j N_{01}}{\alpha \beta_{jn} L_j + d_j N_{01} (\alpha_{jn} - \alpha)} \alpha \beta_{jn} N_j. \tag{7.2.12}$$

The equations, (7.2.11) and (7.2.12), determine the labor and land distribution, L_{js}, L_{jn}, N_{js}, and N_{jn}, within each region as unique functions of N_j and N_{01}. From (7.2.11) and (7.2.12), we see that in order that L_{js}, L_{jn}, N_{js} and N_{jn} are positive, it is necessary to require

$$\beta_0 L_j > d_j N_{01}, \quad \alpha\beta_{jn} L_j + d_j N_{01}(\alpha_{jn} - \alpha) > 0. \tag{7.2.13}$$

In the case of $\beta_0 L_j > d_j N_{01} > 0$, we have

$$\frac{\alpha\beta_{jn} L_j}{d_j} + N_{01}(\alpha_{jn} - \alpha) > N_{01}\left(\frac{\alpha\beta_{jn}}{\beta_0} + \alpha_{jn} - \alpha\right) > 0.$$

Hence, the second inequality is satisfied if $\alpha\beta_{jn} / \beta_0 + \alpha_{jn} > \alpha$. By the definitions of the parameters, we see that it is not strict to require the second inequality to be satisfied. We are only concerned with the first inequality later on.

Substituting (7.2.11) and (7.2.12) into (7.2.4) yields

$$N_1 = \frac{f_1 N}{f_1 + f_2}, \quad N_2 = \frac{f_2 N}{f_1 + f_2} \tag{7.2.14}$$

where we use $N_1 + N_2 = N$ and

$$f_1(N_{01}) \equiv \alpha\beta_{1n} L_1 + d_1 N_{01}(\alpha_{1n} - \alpha),$$

$$f_2(N_{01}) \equiv \{\alpha\beta_{2n} L_2 + d_2 N_{01}(\alpha_{2n} - \alpha)\}\left(\frac{A}{A_0}\right)^{1/\alpha} \frac{\alpha_{1n}}{\alpha_{2n}}. \tag{7.2.15}$$

We now determine N_{01}. Substituting $C_{0j1} = \xi_0 \rho_0 Y_{0j}$ and $C_{j1} = \xi\rho Y_j$ into (7.1.12) yields

$$\frac{d_1 N_0 \xi_0 R_1}{\alpha\eta_0} + \left(N_1 + \frac{N_2}{A}\right)\xi\rho w_1 = F_{1n} \tag{7.2.16}$$

where we use (7.1.3)-(7.1.5), (7.2.2) and (7.2.5). Substituting $R_1 = \alpha_{1n} F_{1n} / L_{1n}$ and $w_1 = \beta_{1n} F_{1n} / L_{1n}$ in (7.1.2) into the above equation, we have

$$\frac{d_1 N_0 \xi_0 \alpha_{1n}}{\alpha \eta_0 L_{1n}} + \left(N_1 + \frac{N_2}{A} \right) \frac{\xi \rho \beta_{1n}}{N_{1n}} = 1. \tag{7.2.18}$$

Substituting (7.2.12) and (7.2.14) into (7.2.18), we get

$$N_{01} = \frac{n_1}{n_2} \tag{7.2.19}$$

where

$$n_1 \equiv \left(\frac{\beta_0}{\xi \rho} - \beta_{1n} + \frac{\alpha_{1n} \beta_{2n}}{\alpha_{2n}} \right) \alpha L_1 +$$
$$\left(\frac{\alpha_0 + \beta_0}{\xi \rho} \xi_0 + \frac{\beta_{2n}}{\alpha_{2n} \rho_0} \right) \frac{d_1 N_0 \alpha_{1n}}{\eta_0},$$
$$n_2 \equiv \left(\frac{2\alpha_{1n}}{\alpha} + \frac{\eta}{\xi} + \frac{\lambda}{\xi} - \frac{\alpha_{1n}}{\alpha_{2n}} \right) \alpha d_1 \qquad (7.2.20$$

in which we require $n_2 > 0$.

We have thus explicitly solved the equilibrium problem. The unique equilibrium values of the 44 variables are given by the following procedure: N_{01} by (7.2.19) $\rightarrow$ $N_{02} = N_0 - N_{01}$ $\rightarrow$ N_j by (7.2.14), $j = 1, 2$ $\rightarrow$ L_{jn} and N_{jn} by (7.2.12) $\rightarrow$ L_{js} and N_{js} by (7.2.12) $\rightarrow$ F_{jn} and F_{js} by (7.1.1) $\rightarrow$ $R_1 = \alpha_{1n} F_{1n} / L_{1n}$ by (7.1.2) $\rightarrow$ $w_1 = \beta_{1n} F_{1n} / L_{1n}$ $\rightarrow$ $p_{1s} = w_1 N_{1s} / \beta F_{1s}$ $\rightarrow$ $p_{2s} = p_{1s} / A_0$ and $w_2 = w_1 / A$ by (7.2.3) $\rightarrow$ $R_2 = \alpha p_{2s} F_{2s} / L_{2s}$ $\rightarrow$ $p_2 = w_2 N_{2n} / \beta_{2n} F_{2n}$ $\rightarrow$ Y_j by (7.1.3) $\rightarrow$ R_0 by (7.1.4) $\rightarrow$ Y_{0j} by (7.1.5) $\rightarrow$ C_{0j1}, C_{0j2}, C_{0js}, C_{j1}, C_{j2}, and C_{js} by (7.1.8) $\rightarrow$ U_j and U_{0j} by (7.1.6).

For the solutions to be meaningful, it is necessary to require N_{01} to satisfy: $\beta_0 L_j > d_j N_{01} > 0$ in (7.2.13) and $N_0 \geq N_{01} \geq 0$. That is

$$\min\left\{N_0,\ \frac{\beta_0 L_j}{d_j}\right\} \geq \frac{n_1}{n_2} \geq 0. \tag{7.2.21}$$

If the above inequalities are not satisfied, the landlord class may be concentrated in a single region.

For illustration, let us examine a special case of $\alpha_{1n} = \alpha_{2n} = \alpha = 1/2$, $\eta_0 = \eta$, $\xi_0 = \xi$, $\lambda_0 = \lambda$, $L_1 = L_2$, and $A = A_0 = 1$. It can be checked that in this case (7.2.21) becomes

$$\min\left\{1,\ \frac{\xi + \lambda}{\eta} + \frac{1}{2}\right\} \geq \frac{\lambda + \eta/2 - \xi}{\eta} \geq 0. \tag{7.2.22}$$

The above inequalities simply imply that if the two classes strongly prefer either of the two region-specified goods, then the landlord class may be concentrated in a single region. This is obviously intuitively acceptable under the presumed conditions. From the definitions of the parameters, we see that the inequalities, (7.2.21), are determined by the population and preference structures of the two classes, and the land sizes, amenity levels, and technological structures of the two regions.

Summarizing the above discussion, we have the following proposition.

Proposition 7.2.1.
If (7.2.21) holds, the economic system has a unique equilibrium.

7.3 The Farmers and the Economic Structure

This section examines the effects of changes in the number, N, of the farmer class on the economic structure. Taking derivatives of (7.2.19) and $N_{02} = N_0 - N_{01}$ with respect to N yields

$$\frac{dN_{01}}{dA} = \frac{dN_{02}}{dA} = 0. \tag{7.3.1}$$

An increase in the farmer class population has no impact on the regional population distribution of the landlord class.

By (7.2.14), (7.2.12) and (7.2.11), we get

$$\frac{1}{N_1}\frac{dN_1}{dA} = \frac{1}{N_{js}}\frac{dN_{js}}{dA} = \frac{1}{N_{jn}}\frac{dN_{jn}}{dA} = \frac{1}{N} > 0, \quad \frac{dL_{js}}{dA} = \frac{dN_{jn}}{dA} = 0. \tag{7.3.2}$$

The labor force, N_j, of the two regions and labor inputs, N_{js} and N_{jn} of each region's two sectors are increased, and the land distribution, L_{js} and L_{jn}, within each region is not affected. Taking derivatives of (7.1.1), we obtain

$$\frac{1}{F_{jn}}\frac{dF_{jn}}{dA} = \frac{\beta_{jn}}{N} > 0, \quad \frac{1}{F_{js}}\frac{dF_{js}}{dA} = \frac{\beta_{js}}{N} > 0. \tag{7.3.3}$$

The output levels, F_{jn} and F_{js}, of the two sectors in each region are increased. By (7.1.2), we directly get the impact on region 1's prices as follows

$$\frac{1}{R_1}\frac{dR_1}{dA} = \frac{\beta_{jn}}{N} > 0, \quad \frac{1}{w_1}\frac{dw_1}{dA} = -\frac{\alpha_{1n}}{N} < 0, \quad \frac{1}{p_{1s}}\frac{dp_{1s}}{dA} = \frac{\beta_{1n} - \beta}{N}. \tag{7.3.4}$$

Region 1's land rent (wage rate), R_1 (w_1), is increased (reduced) and region 1's services price, p_{1s}, is increased (reduced) in the case of $\beta_{1n} > (<) \beta_{1s}$. Using (7.3.4), (7.2.2), (7.2.5) and $p_2 = R_2 L_{2n} / \alpha_{2n} F_{2n}$, we get

$$\frac{1}{R_2}\frac{dR_2}{dA} = \frac{1}{R_1}\frac{dR_1}{dA} > 0, \quad \frac{1}{w_2}\frac{dw_2}{dA} = \frac{1}{w_1}\frac{dw_1}{dA} < 0,$$
$$\frac{1}{p_{2s}}\frac{dp_{2s}}{dA} = \frac{1}{p_{1s}}\frac{dp_{1s}}{dA}, \quad \frac{1}{p_{2s}}\frac{dp_{2s}}{dA} = \frac{\beta_{1n} - \beta_{2n}}{N}. \tag{7.3.5}$$

Region 2's land rent (wage rate), R_2 (w_2), is increased (reduced), region 2's services price, p_{2s}, is increased (reduced) in the case of $\beta_{1n} > (<) \beta$ and the price, p_2, region 2's region-specified good is increased (decreased) in the case of $\beta_{1n} > (<) \beta_{2n}$.

By (7.1.3) and (7.1.5), we have

$$\frac{1}{Y_{0j}}\frac{dY_{0j}}{dA}=\frac{\beta_{1n}}{N}>0,\ \ \frac{1}{Y_j}\frac{dY_j}{dA}=\frac{\beta_{1n}}{N}>0. \tag{7.3.6}$$

The net incomes, Y_{0j} and Y_j, of each class in the two regions are increased.

Taking derivatives of (7.1.8) with respect to N and using the above results, we get

$$\frac{1}{\beta_{1n}C_{0j1}}\frac{dC_{0j1}}{dA}=\frac{1}{\beta_{2n}C_{0j2}}\frac{dC_{0j2}}{dA}=\frac{1}{\beta C_{0j2}}\frac{dC_{0j2}}{dA}=\frac{1}{N}>0,$$
$$\frac{1}{\beta_{1n}C_{j1}}\frac{dC_{j1}}{dA}=\frac{1}{\beta_{2n}C_{j2}}\frac{dC_{j2}}{dA}=\frac{1}{\beta C_{j2}}\frac{dC_{j2}}{dA}=\frac{1}{N}>0. \tag{7.3.7}$$

The effects on the net income and consumption levels per capita of the landlord class are given as follows

$$\frac{1}{\beta_{1n}y_{0j}}\frac{dy_{0j}}{dA}=\frac{1}{\beta_{2n}c_{0j2}}\frac{dc_{0j2}}{dA}=\frac{1}{\beta c_{0js}}\frac{dc_{0js}}{dA}=\frac{1}{N}>0 \tag{7.3.8}$$

where we use $y_{0j} \equiv Y_{0j}/N_{0j}$ and $c_{0jk} \equiv C_{0jk}/N_{0j}$.

The effects on the net income and consumption levels per capita of the farmer class are given as follows

$$\frac{1}{\beta_{1n}y_j}\frac{dy_j}{dA}=\frac{1}{\beta_{2n}c_{j2}}\frac{dc_{j2}}{dA}=\frac{1}{\beta c_{js}}\frac{dc_{js}}{dA}=-\frac{1}{N}<0 \tag{7.3.9}$$

where we use $y_j \equiv Y_j/N_j$ and $c_{jk} \equiv C_{jk}/N_j$. Summarizing the discussion in this section, we get the following proposition.

Proposition 7.3.1.
An increase on the number N of the farmer class has the following impact on the economic structure:
(1) on the price structure: The farmers' wage rates, w_j, in the two regions are reduced, the levels, R_1 and R_2, of land rent in the two regions and the income per capita, R_0, of the landlord class are increased, the services prices, p_{js}, in the two regions are increased (reduced) in the case of $\beta_{1n} > (<) \beta$, and the price of region

2 's region-specified commodity is increased (reduced) in the case of $\beta_{1n} > (<)\ \beta_{2n}$; (2) on the aggregated variables of the farmer class: The numbers, N_j, the net incomes, Y_j, and the consumption levels, C_{js}, C_{j1} and C_{j2}, of services and the two region-specified goods in the two regions are increased; (3) on the variables per capita of the farmer class: The net incomes, y_j, and the consumption levels, c_{js}, c_{j1} and c_{j2}, of services and the two region-specified goods in the two regions are reduced; (4) on the aggregated variables of the landlord class: The regional distribution, N_{0j}, of the landlord class population is not affected, and the net incomes, Y_{0j}, the consumption levels, C_{0js}, C_{0j1} and C_{0j2}, of services and the two region-specified goods the in the two regions are increased; (5) on the variables per capita of the landlord class: The net incomes, y_{0j}, and the consumption levels, c_{0js}, c_{0j1} and c_{0j2}, of services and the two region-specified goods in the two regions are increased; and (6) on the sector distribution of labor and land: The land distributions, L_{js} and L_{jn}, between the two sectors in the two regions are not affected, and the labor numbers, N_{js} and N_{jn}, employed by the two sectors in the two regions are increased.

7.4 The Amenity of Landlords and Economic Structure

This section examines the effects of changes in region 1 's amenity level, A_{01}, of the landlord class on the economic structure. Taking derivatives of (7.2.19) and $N_{02} = N_0 - N_{01}$ with respect to A_{01} yields

$$\frac{1}{N_{01}}\frac{dN_{01}}{dA_{01}} = \frac{\beta_0 / \rho\xi - \beta_{1n} + \alpha_{1n}\beta_{2n} / \alpha_{2n}}{L_1 + L_2 A^{\beta/\alpha} / A_0^{1/\alpha}} \frac{\beta_0 L_1 L_2 A^{\beta/\alpha}}{A_0^{1/\alpha} A_{01}\eta_0 n_1} > 0,$$

$$\frac{dN_{02}}{dA_{01}} = -\frac{dN_{01}}{dA_{01}} < 0. \tag{7.4.1}$$

As the landlord class' amenity level for living in region 1 is increased, the number of region 1 's (2 's) landlord class is increased (reduced).

Taking derivatives of (7.2.14) with respect to A_{01} yields

$$\frac{1}{N_1}\frac{dN_1}{dA_{01}} = \frac{(1/f_1)df_1/dA_{01} - (1/f_2)df_2/dA_{01}}{f_1 + f_2} f_2, \quad \frac{dN_2}{dA_{01}} = -\frac{dN_1}{dA_{01}}. \tag{7.4.2}$$

where f_1 and f_2 are defined in (7.2.15). It is difficult to explicitly judge the sign of $(1/f_1)df_1/dA_{01} - (1/f_2)df_2/dA_{01}$. It can be seen that the term may be either positive or negative. As the distribution is affected by the two classes' preferences, the two regions' technologies, the ownership structure, and many other factors, this conclusion is expectable. For illustration, we assume $\alpha = \alpha_{1n} = \alpha_{2n}$ in the remainder of this section. In this case, we rewrite (7.4.2) as follows

$$\frac{1}{N_1}\frac{dN_1}{dA_{01}} = \frac{(A/A_0)^{1/\alpha} L_2}{L_1 + L_2 (A/A_0)^{1/\alpha}} \frac{1 - \eta\rho}{\alpha\eta_0 A_{01}}, \quad \frac{dN_2}{dA_{01}} = -\frac{dN_1}{dA_{01}}. \tag{7.4.3}$$

The number of the farmer class in region 1 (region 2) is increased (reduced). By (7.2.11), (7.2.12) and (7.1.1), we obtain

$$\frac{1}{L_{js}}\frac{dL_{js}}{dA_{01}} = \frac{1}{d_j N_{01} + \alpha_0 L_j}\frac{d(d_j N_{01})}{dA_{01}},$$

$$\frac{1}{N_{js}}\frac{dN_{js}}{dA_{01}} = \frac{1}{L_{js}}\frac{dL_{js}}{dA_{01}} + \frac{1}{N_j}\frac{dN_j}{dA_{01}},$$

$$\frac{1}{L_{jn}}\frac{dL_{jn}}{dA_{01}} = \frac{1}{d_j N_{01} - \beta_0 L_j}\frac{d(d_j N_{01})}{dA_{01}},$$

$$\frac{1}{N_{jn}}\frac{dN_{jn}}{dA_{01}} = \frac{1}{L_{jn}}\frac{dL_{jn}}{dA_{01}} + \frac{1}{N_j}\frac{dN_j}{dA_{01}},$$

$$\frac{1}{F_{js}}\frac{dF_{js}}{dA_{01}} = \frac{1}{L_{js}}\frac{dL_{js}}{dA_{01}} + \frac{\beta}{N_j}\frac{dN_j}{dA_{01}},$$

$$\frac{1}{F_{jn}}\frac{dF_{jn}}{dA_{01}} = \frac{1}{L_{jn}}\frac{dL_{jn}}{dA_{01}} + \frac{\beta}{N_j}\frac{dN_j}{dA_{01}}, \quad j = 1, 2 \tag{7.4.4}$$

where

$$\frac{1}{d_1 N_{01}} \frac{d(d_1 N_{01})}{dA_{01}} =$$

$$\frac{(A-1)\beta_0 - \alpha \eta \rho A\left(L_1 + L_2 A^{\beta/\alpha} / A_0^{1/\alpha}\right)}{A\left(L_1 + L_2 A^{\beta/\alpha} / A_0^{1/\alpha}\right)^2} \frac{L_2 (A / A_0)^{1/\alpha}}{\alpha \eta_0 A_{01}},$$

$$\frac{1}{d_2 N_{01}} \frac{d(d_2 N_{01})}{dA_{01}} = \frac{1}{d_1 N_{01}} \frac{d(d_1 N_{01})}{dA_{01}} + \frac{\beta_0}{\alpha \eta_0 A_{01}}. \tag{7.4.5}$$

The effects are determined by the signs of $d(d_j N_{01}) / dA_{01}$ and dN_2 / dA_{01}. Even in this simplified case, it is difficult to explicitly judge the impact of changes in A_{01} on the labor and land distribution within each region and output levels. Similarly to the preceding section, we may analyze the effects of changes in A_{01} on the remaining variables in the system. Because it is difficult to explicitly interpret the results, we will not discuss further.

7.5 On Classes and Regional Economic Structure

This chapter proposed a two-region and two-class equilibrium model of perfect competition. We classified the population into two, farmer and landlord, classes. The farmer class produces goods and services and gets its wage income. The landlord class gets incomes from its land revenue. Each region is assumed to produce two kinds of production, services and region-specified goods. Production is assumed to employ two kinds of inputs, land and labor. Services are consumed by the producer region and two region-specified goods are freely mobile between regions. The model determines a unique equilibrium of the economy, with given preferences, regional amenities, technologies and resources (of land and labor). We also examined the effects of changes in the farmer class population and region 1's amenity level for the landlord class on equilibrium structure of the economy.

This chapter is concerned only with a perfectly competitive two-region (agricultural) economy with constant return-to-scale effects in all economic production. It is important to extend the model to economies with return to scale effects. This can be carried out, for instance, by introducing infrastructures, capital, population or knowledge dynamics into the system. The institutional structure specified in this chapter is simplified. Land property ownership is a difficult issue. Even in agricultural economies land ownership and taxation structures may exhibit a large variety in different cultures. There are obviously many ways to make regional amenities as endogenous variables. For instance, it is reasonable to introduce (regional or national) public sectors to supply public goods to regions. Regional public sectors may interact with the private economy in many ways. In this chapter, we omit transportation costs, which are important for determining regional division of labor (e.g., Batten, 1982, Beckmann, 1969, 1976, 1990, Beckmann and Puu, 1985). This assumption may be relaxed by introducing a special service - transportation - sector, which is specified in transporting goods between the two regions. With road systems and other transportation infrastructures between the two regions fixed, we may use labor force and volumes of goods to be transported as inputs in the transportation sector's "production function". Accordingly, transportation sector may be treated as other production sectors in our system.

8 Regional Growth with Economic Structure

The previous chapters of regional dynamics are concerned with one-sector economies. But to explain economic mechanism of division of labor and competitive equilibrium in labor market, the one-sector model is not sufficient to provide insights into dynamic processes of division of labor and interdependence between division of labor, knowledge and efficiency. It is necessary to extend the one-sector economy into a multi-sector economy. This chapter is to provide a theoretical framework for studying the interdependence among the division of labor, growth of population, wealth accumulation, income distribution, and prices and profits determination within a multi-regional context.

It may be argued that the main task of developmental economics is to explain how structural changes occur in economic systems (Rostow, 1960, Kuznets, 1963, 1966, Lewis, 1955). Since the pioneering works of Leontief (Leontief, 1941), numerous theoretical studies on economic structure and dynamics have been published (Sraffa, 1960, Nikaido, 1968, Morishima, 1964, 1969, Pasinetti, 1981). It is well observed that changes in the absolute levels of macro-economic variables such as gross national product, total consumption and total investment are associated with changes in their composition in industrializing processes of many economies. Yet, though the input-output system has proved to be very effective for analyzing economic structure with complicated linkages among various sectors in multi-regional or multi-national systems, it is argued that it is often analytically difficult to introduce endogenous behavior of households in an effective way in this framework. But the role of consumption demand in industrial revolutions and its importance to growth and structural change have long recognized by economists. It has been argued that Engel effects not only cause a shift in the industrial origin of production, but also induce high levels of productivity and output (Kuznets, 1966). Nevertheless, theoretical analysis of this role has been rather limited in dynamic theory. Many of multi-sector economic growth models have omitted a consideration of the role of consumption demand structure. For instance, in the Uzawa's two-sector growth model and its extensions the economic structure is not designed to confront issues of demand, since only one consumption sector is postulated. Although the role of consumption demand is explicitly recognized in the neo-Keynesian literature, the level of aggregation in most of these models is such that there is little scope for changes in the composition of consumption to influence the pattern of growth and structural change.

In the past two decades, analysts have become increasingly aware of the need to improve the specification of household behavior in the input-output models. There are some multi-sector growth models which are concerned with capital accumulation in the neoclassical growth literature (see the appendix to this chapter). Pasinetti (1993) re-addressed issues related to structural dynamics. He studies the role of technical progress in evolution; one of problems of his approach is that, like most of recent growth models with endogenous knowledge, capital accumulation is neglected. This chapter proposes a new framework for examining structural issues within an interregional context. To make an initial attempt to synthesize ideas in location theory, urban economics, economic geography, equilibrium theory and neoclassical growth theory, we construct a dynamic model to describe the interaction between capital accumulation, the regional distribution of capital and labor, the division of labor, the capital distribution in each region, land rents, service prices and commodity prices.

The model includes three spatial factors: land, amenities and services. A regional analysis must treat services differently from commodity goods since many services provided by schools, hospitals and restaurants must be consumed where they are supplied. The spatial character of services implies that service prices will be spatially varied. Amenities such as climatic conditions and historical buildings are location-dependent and have an important impact upon residential location. The remainder of the paper is organized as follows. Section 8.1 defines the basic model. Section 8.2 analyzes properties of the dynamic system. Section 8.3 examines effects of changes in the amenity level on the equilibrium structure of economic geography. Section 8.4 concludes the chapter. The appendix proves the results in Section 8.2. This chapter is based on Zhang (1998a).

8.1 Regional Growth with Structural Change

The economic system consists of two regions, indexed by 1 and 2, respectively. Each region has two sectors: an industrial and a service sector. Services are regionally specified and we disregard the possibility that one region's services, such as tourism and banking, are used by the other. Service and industrial production occurs via the combination of labor and capital under constant returns and perfect competition. It is assumed that the two regions' industrial product is qualitatively homogenous and is either consumed or invested. People are free to choose their residential location. It is assumed that people live in the same region where they work. The land is only used for housing. We select the industrial commodity to serve as numeraire, with all the other prices being measured relative to its price. We disregard the transportation cost of commodities between and within regions. The assumption of a zero transportation cost for commodities implies price equality for the commodity between the two regions. As will become clear later on, although it is not conceptually difficult to introduce a transportation cost function and to provide balance conditions for demand and supply and for price equalization conditions with transportation costs, the problem becomes too complicated in analytical terms. We will discuss the necessity for and the possible ways of relaxing the assumption of zero transportation costs. Let L_j and N stand for region j's fixed land and the fixed population of the economy. We introduce

i, s — the subscript indices for industry and services, respectively;

$F_{ij}(t)$ and $F_{sj}(t)$ — the output levels of region j's industrial and service sectors at time t;

$N_{ij}(t)$ and $N_{sj}(t)$ — the labor force employed by region j's industrial and service sectors in region j;

$K_{ij}(t)$ and $K_{sj}(t)$ — the capital stocks employed by region j's industrial and service sectors;

$K(t)$ — the total capital stocks of the country;

$k_j(t)$ — the capital stocks owned by per worker in region j;

$r(t)$ — the rate of interest; and

$w_j(t)$, $p_j(t)$ and $R_j(t)$ — region j's wage rate, the price of services, and the land rent.

Behavior of producers

We specify the regions' commodity production functions as follows

$$F_{ij}(t) = K_{ij}^{\alpha}(t)N_{ij}^{\beta}(t), \quad \alpha, \beta > 0, \quad \alpha + \beta = 1. \tag{8.1.1}$$

The marginal conditions are given by

$$r(t) = \frac{\alpha F_{ij}}{K_{ij}}, \quad w_j(t) = \frac{\beta F_{ij}}{N_{ij}}. \tag{8.1.2}$$

From (8.1.1) and (8.1.2), we directly have

$$\frac{K_{i1}}{K_{i2}} = \frac{N_{i1}}{N_{i2}}, \quad w(t) \equiv w_1(t) = w_2(t). \tag{8.1.3}$$

The two regions' wage rates are equal. This is due to the assumptions that the two regions have identical forms of production functions as in (8.1.1) and that the interest rate is identical across the whole economy. The service sectors' production functions and the marginal conditions are given by

$$F_{sj}(t) = K_{sj}^{\alpha_j}(t)N_{sj}^{\beta_j}(t), \ \ \alpha_j, \beta_j > 0, \ \ \alpha_j + \beta_j = 1,$$

$$r(t) = \frac{\alpha_j p_j F_{sj}}{K_{sj}}, \ \ w_j(t) = \frac{\beta_j p_j F_{sj}}{N_{sj}}. \tag{8.1.4}$$

Behavior of households

Each worker earns income from land ownership, capital ownership and wages. We specify land revenue as follows. We assume that each worker owns L_j / N amount of land in region j and that it is impossible to sell land but it is possible to rent one's own land to others. This assumption implies that a worker receives land revenue, R_0, irrespective of whether his dwelling site is located in region 1 or region 2. It should be remarked that if we did allow land to be sold in land markets, each person would buy the same amount of the land in each region due to the assumption of perfect competition, perfect information and preferences. With the assumed land ownership structure, we get the following land revenue per worker

$$R_0 = \frac{L_1 R_1 + L_2 R_2}{N}. \tag{8.1.5}$$

The net income per worker, $y_j(t)$, in region j consists of his wage income, $w_j(t)$, land revenue, $R_0(t)$, and interest payment, $r(t)k_j(t)$, i.e.

$$y_j(t) = R_0 + w + rk_j. \tag{8.1.6}$$

It is assumed that a typical worker's utility level is dependent on his consumption levels, $c_{js}(t)$ and $c_{ji}(t)$, of services and industrial goods, his lot size, $l_j(t)$, and the level of his wealth, $k_j(t) + s_j(t) - \delta_k k_j(t)$, where $s_j(t)$ is the savings made by a worker in region j at time t and δ_k is the fixed depreciation rate of capital. The utility function is specified as follows

$$U_j(t) = A_j l_j^{\eta} c_{js}^{\gamma} c_{ji}^{\xi} \left(k_j + s_j - \delta_k k_j\right)^{\lambda} \tag{8.1.7}$$

where U_j represents the utility levels of region j's workers, and the positive parameters are the propensities to consume housing, to consume services, to consume goods, and to hold wealth respectively. In (8.1.7), A_j is called region j's amenity level. It is not difficult to propose some reasonable assumptions to represent amenities as endogenous variables in the form of

$$\frac{dA_j}{dt} = D_j(A_j ...)$$

where D_j are certain functions of the endogenous variables. If we specify D_j properly, we can take account of endogenous amenities within our analytical framework.

The financial budget constraints are given by

$$R_j l_j + p_j c_{js} + c_{ji} + s_j = y_j . \tag{8.1.8}$$

A typical household maximizes (8.1.7) subject to (8.1.8). The optimal problem has the following solutions

$$R_j l_j = \eta \Omega_j , \quad p_j c_{js} = \gamma \Omega_j , \quad c_{ji} = \xi \Omega_j , \quad s_j = \lambda \Omega_j - (1 - \delta_k) k_j \tag{8.1.9}$$

where

$$\Omega_j \equiv \rho y_j + \delta k_j , \quad \delta \equiv (1 - \delta_k) \rho , \quad \rho \equiv \frac{1}{\eta + \gamma + \xi + \lambda} .$$

According to the definitions of k_j and s_j, we have

$$\frac{dk_j}{dt} = s_j - \delta_k k_j .$$

Substituting s_j in (8.1.9) into these two equations yields

$$\frac{dk_j}{dt} = \lambda \Omega_j - k_j , \quad j = 1, 2 . \tag{8.1.10}$$

By the definitions of $\Omega_j(t)$ and $w_1(t) = w_2(t)$, $\Omega_1(t) = \Omega_2(t)$ holds if $k_1(t) = k_2(t)$. From this and (8.1.11), if $k_1(0) = k_2(0)$, then $k_1(t) = k_2(t)$. We assume $k_1(0) = k_2(0)$. This guarantees

$$k_1(t) = k_2(t), \quad \Omega(t) \equiv \Omega_1(t) = \Omega_2(t) . \tag{8.1.11}$$

The households in the two regions have the same degree of wealth and get the same level of net income. Using $K(t) = k_1 N_1 + k_2 N_2$, $dN / dt = 0$ and (8.1.11), we get

$$\frac{dK}{dt} = \lambda N \Omega - K . \tag{8.1.12}$$

Everyone's utility level of people should be equal, irrespective of the region in which they live, i.e.

$$U_1(t) = U_2(t) . \tag{8.1.13}$$

Let $C_{ji}(t)$ and $C_{js}(t)$ respectively stand for the total consumption levels of the industrial commodity and services by region j. According to our definitions, we have

$$C_{ji}(t) = c_{ji}(t) N_j(t), \quad C_{js}(t) = c_{js}(t) N_j(t) .$$

The balance of the demand for and the supply of services and industrial commodities is give

$$C_{js}(t) = F_{js}(t) , \tag{8.1.14}$$

$$\sum_j \{ C_{js}(t) + s_j(t) N_j(t) \} = \sum_j F_{ij}(t) . \tag{8.1.15}$$

It is assumed that capital, labor and land are fully employed in the regions, i.e.

$$\sum_j \left(K_{ij} + K_{sj} \right) = K, \quad N_{ij} + N_{sj} = N_j, \quad N_1 + N_2 = N,$$

$$l_j N_j = L_j . \tag{8.1.16}$$

We built the model. The system has 41 independent variables, N_{sj}, N_{ij}, K_{sj}, N_{ij}, F_{sj}, F_{ij}, C_{js}, C_{ji}, Y_j, l_j, c_{js}, c_{ji}, y_j, U_j, N_j, s_j, p_j, w_j, R_j ($j = 1, 2$), K, r, and R_0. The system contains the same number of independent equations.

8.2 The Regional Equilibrium Structure

First, we show that the dynamic behavior of the system can be described by a single differential equation. Then, we provide the conditions for the existence of equilibria and stability. The following lemma is proved in the appendix.

Lemma 8.2.1

Any variable in the dynamic system can be expressed as functions of $K(t)$ at any point of time. In other words, if we know the value of $K(t)$ at time t, then we can determine all the variables at t. If

$$\min\{\alpha\theta(a_1 + a_2), \beta\theta(a_1 + a_2)\} > a \equiv a_2 - a_1$$

then any variable is determined as a unique function of $K(t)$.

The above lemma guarantees that any variable in the system can be expressed as functions of K at any point of time. The requirement,

$$\min\{\alpha\theta(a_1 + a_2), \beta\theta(a_1 + a_2)\} > a \ (> 0)$$

is guaranteed if a is small. We explicitly provide the procedure to determine values of all the variables for a given $K(t)$ at any point of time. We now show how we determine the motion of $K(t)$. By Lemma 8.2.1 and (8.1.12), we see that the economic dynamics are expressed as one-dimensional dynamics of K as follows

$$\frac{dK}{dt} = \lambda K\Omega(K) - K \tag{8.2.1}$$

in which $\Omega(K)$ is a function of K determined in the appendix. The motion of $K(t)$ is determined by (8.2.1). The other variables are determined as functions of $K(t)$ through the assumed condition of perfect competition among the economic agents. It is sufficient to examine the properties of (8.2.1).

Equilibrium of (8.2.1) is given by

$$\lambda K\Omega(K) = K . \tag{8.2.2}$$

By (8.2.2), (8.A.1.11) and (8.A.1.12), we solve

$$K = \lambda N w K_0 \tag{8.2.3}$$

in which

$$K_0 \equiv \frac{1}{\beta\xi + \beta\lambda\delta_k + (b_1\Lambda + b_2)\gamma / \Lambda_1}$$

where Λ is defined in the appendix and $\Lambda_1(\Lambda) \equiv \Lambda + 1$. Substituting (8.A.1.13) and (8.A.1.3) into (8.A.1.2), we have

$$\Phi(\Lambda) \equiv \frac{\alpha\lambda}{b_0^\beta \Lambda^{\beta\theta}} - \frac{a_1\Lambda + a_2}{\Lambda_1}\gamma - \alpha(\xi + \delta_k\lambda) = 0 \qquad (8.2.4)$$

where we use $r = \alpha / b_0^\beta \Lambda^{\beta\theta}$ given in the appendix. We have

$$\Phi(0) > 0, \ \ \Phi(+\infty) < 0, \ \ \Phi_1 \equiv \Lambda_1^2 \frac{d\Phi}{d\Lambda} = a\gamma - \frac{\alpha\beta\lambda\theta\Lambda_1^2}{b_0^\beta \Lambda^{\beta\theta+1}}. \qquad (8.2.5)$$

As $\Phi(0) > 0$ and $\Phi(+\infty) < 0$, $\Phi(\Lambda) = 0$ has at least one positive solution. In the case of $\beta\theta < 1$, i.e.

$$\frac{\beta\eta}{\lambda} < a_2 - a_1, \ \ \Phi_1(0) < 0, \ \ \Phi_1(+\infty) < 0$$

and $\Phi_1^{'}(\Lambda) = 0$ has a unique solution

$$\Lambda_0 \equiv \frac{\beta\theta + 1}{\beta\theta - 1} > 0.$$

It can be seen that $\Phi_1(\Lambda)$ is increasing in the interval $(0, \Lambda_0)$ and decreasing in the interval $(\Lambda_0, +\infty)$. This implies that $\Phi'(\Lambda) = 0$ has two positive solutions if there is a positive Λ at which $\Phi'(\Lambda) > 0$. Hence, in the case of $\beta\theta < 1$, $\Phi(\Lambda) = 0$ may have multiple solutions. In the case of $\beta\theta > 1$, i.e.

$$\frac{\beta\eta}{\lambda} > a, \ \ \Phi_1(0) < 0, \ \ \Phi_1(+\infty) > 0$$

and $\Phi'(\Lambda) > 0$ for any $\Lambda > 0$. This implies that $\Phi'(\Lambda) = 0$ has a unique positive solution. As $\Phi(0) > 0$ and $\Phi(+\infty) < 0$ and $\Phi(\Lambda)$ is continuous, if $\Phi(\Lambda) = 0$ has multiple solutions, $\Phi'(\Lambda) = 0$ also has multiple solutions. This implies that $\Phi(\Lambda) = 0$ has a unique solution. As $\beta\eta > a\lambda$ implies that the difference, $a = a_2 - a_1$, in the elasticity of capital of the two regions' services production is small, we see that the requirement of $\beta\theta > 1$ is quite acceptable. For any given equilibrium value, Λ, K is uniquely determined by (8.A.1.3) and equilibrium values of the other variables are uniquely given by the procedure in the appendix.

By (8.A.1.1), (8.A.1.13), $r = \alpha / b_0^{\beta} \Lambda^{\beta\theta}$ and $w = \beta b_0^{\beta} \Lambda^{\alpha\theta}$, we calculate the eigenvalue

$$\phi = \lambda N \frac{d\Omega}{dK} - 1 = \frac{\alpha w - \beta r}{1 - \eta\rho} \frac{\lambda\theta\rho}{\Lambda} \frac{d\Lambda}{dK} + \frac{\lambda\rho r + \delta\lambda - 1 + \eta\rho}{1 - \eta\rho} \tag{8.2.6}$$

where by (8.A.1.14) we have

$$\Phi^* \frac{d\Lambda}{dK} = - \left\{ \beta\xi + \beta\lambda + \frac{b_1\Lambda + b_2}{\Lambda_1} \gamma \right\} \frac{\alpha}{b_0^{\beta} \Lambda^{\beta\theta}} + \left\{ \beta - \frac{b_1\Lambda + b_2}{\Lambda_1} \right\} \frac{\delta\gamma}{\rho} \tag{8.2.7}$$

in which Φ^* is defined by (8.A.1.15) and is evaluated at the equilibrium. As it is not easy to interpret the sign of ϕ, we will not examine the stability conditions further. Summarizing the discussion in this section, we have the following proposition.

Proposition 8.2.1.
Assume $a > 0$. The dynamic system has at least one equilibrium. If $\beta\eta > a\lambda$, the system has a unique equilibrium. An equilibrium is stable (unstable) if $\phi < (>)\, 0$.

The remainder of this chapter requires

$$\beta\eta > a\lambda .$$

We will examine the effects of changes in some parameters on the economic structure. By (8.2.4), (8.2.5) and $\beta\theta > 1$, we have

$$\Lambda_1^2 \frac{d\Phi}{d\Lambda} = (\Lambda - \beta\theta\Lambda - \beta\theta)\frac{a_2\gamma}{\Lambda} - a_1\gamma - a_1\gamma\beta\theta\Lambda_1$$
$$- \frac{\xi + \delta_k\lambda}{\Lambda}\alpha\beta\theta\Lambda_1^2 < 0. \tag{8.2.8}$$

As shown below, $\Phi' < 0$ is important in comparative static analysis.

8.3 The Regional Amenity and the Economic Structure

This section is concerned with the effects of changes in region 1's amenity level, A_1, on the equilibrium structure. Taking derivatives of (8.2.4)

$$N_1 = \frac{\Lambda N}{1 + \Lambda}, \quad N_2 = \frac{N}{1 + \Lambda}$$

with respect to A_1, we get

$$\Phi' \frac{d\Lambda}{dA_1} = \frac{\alpha\beta\theta\lambda}{\eta b_0^\beta A_1 \Lambda^{\beta\theta}} > 0, \quad \frac{1}{N_1}\frac{dN_1}{dA_1} = \frac{1}{\Lambda\Lambda_1}\frac{d\Lambda}{dA_1} < 0,$$
$$\frac{1}{N_2}\frac{dN_2}{dA_1} = -\frac{1}{\Lambda_1}\frac{d\Lambda}{dA_1} > 0 \tag{8.3.1}$$

where $\Phi' < 0$ is defined in (8.2.8).

As region 1's amenity is improved, less people will live in region 1. Some of region 1's population will migrate to region 2. As an improvement in region 1's amenity makes region 1 more attractive at the initial stage, more people will move to region 1. This tends to increase region 1's land rent and the price services at the beginning. On the other hand, as the two regions' service sectors have different technologies ($b_1 > b_2$) and the wage rate is the same in both regions, region 2 may become more attractive as people migrate away. The long-run balance of multiple forces results in a decline of region 1's population as region 1's amenity is improved (with its technology fixed). In reality, as a region's amenity is improved, its population tends to increase. In our model, the amenity factor is assumed to be a constant multiplicative factor of the utility function and to be independent of any other variables. In practical terms, amenity has a crowding effect. Since there is no crowding effect in our model, to equalize the utility level of both regions the region-specific prices (of land rents and services) have to increase, which leads to the decrease of the population in region 1.

By (8.1.4) and (8.1.8), we have

$$p_1 = \frac{w}{b_1}\left(\frac{a_2 b_1}{a_1 b_2 b \Lambda^\theta}\right)^{a_1}, \quad p_2 = \frac{w}{b_2}\left(\frac{1}{b\Lambda^\theta}\right)^{a_2}. \tag{8.3.2}$$

By $r = \alpha / b_0^\beta \Lambda^{\beta\theta}$, $w = \beta b_0^\beta \Lambda^{\alpha\theta}$ and (8.3.2), we have

$$\frac{1}{r}\frac{dr}{dA_1} = -\left(\frac{1}{\Lambda}\frac{d\Lambda}{dA_1} + \frac{1}{\eta A_1}\right)\beta\theta > 0,$$
$$\frac{1}{w}\frac{dw}{dA_1} = \left(\frac{1}{\Lambda}\frac{d\Lambda}{dA_1} + \frac{1}{\eta A_1}\right)\alpha\theta < 0,$$
$$\frac{1}{p_j}\frac{dp_j}{dA_1} = \left(\frac{1}{\Lambda}\frac{d\Lambda}{dA_1} + \frac{1}{\eta A_1}\right)(\alpha - a_j)\theta \tag{8.3.3}$$

where we use

$$\frac{1}{\Lambda}\frac{d\Lambda}{dA_1} + \frac{1}{\eta A_1} = \frac{\alpha\gamma}{\eta A_1 \Lambda_1^2 \Phi'} < 0$$

which is obtained from (8.3.1), (8.2.23) and (8.2.4). As region 1's amenity is improved, the rate of interest, r, is increased, the wage rate, w, is decreased and the price of region j's services, p_j, is increased (decreased) in the case of $a_j > \alpha$. By (8.2.2) and (8.2.3), we have

$$\frac{1}{K}\frac{dK}{dA_1} = \frac{1}{\Omega}\frac{d\Omega}{dA_1} = (b_1 + b_2\Lambda)\frac{a\theta\gamma^2 K_0}{\eta A_1 \Phi' \Lambda_1^3} < 0 \tag{8.3.4}$$

which is obtained from (8.3.3), (8.2.8) and (8.A.1.1). The total capital stocks, K, are reduced as region 1's amenity is improved. In the remainder of this section we require $\delta_k = 0$. By (8.A.1.11), (8.2.2) and (8.2.3), we have

$$N_{s1} = \frac{b_1 \gamma a K_0 N}{1 + \Lambda}.$$

By this equation, (8.A.1.6), (8.A.1.8), $N_{ij} = N_j - N_{sj}$ and (8.A.1.10), we get

$$\frac{\Lambda_1}{N_{s1}}\frac{dN_{s1}}{dA_1} = K_0 \frac{\beta\xi + \gamma b_2}{\Lambda}\frac{d\Lambda}{dA_1} < 0,$$

$$\frac{\Lambda_1}{N_{s2}}\frac{dN_{s1}}{dA_1} = - K_0(\beta\xi + \gamma b_1)\frac{d\Lambda}{dA_1} > 0,$$

$$\frac{n_1\Lambda_1^2}{N_{i1}}\frac{dN_{i1}}{dA_1} = \left(\beta\xi\Lambda^2 + 2\beta\xi\Lambda - a\gamma\right)\frac{\beta\xi + \gamma b_2}{\Lambda}\frac{d\Lambda}{dA_1} < 0,$$

$$\frac{n_2\Lambda_1^2}{N_{i2}}\frac{dN_{i2}}{dA_1} = - \frac{\{2\beta\xi + (\beta\xi + a\gamma)\Lambda\}(\beta\xi + \gamma b_1)\Lambda + \beta^2\xi^2 + (\beta\xi - a\gamma)\gamma b_2}{\Lambda}\frac{d\Lambda}{dA_1} \tag{8.3.5}$$

where

$$n_j \equiv \frac{\Lambda_1^2 N_{ij}}{N\Lambda K_0^2} > 0.$$

The number of workers employed by region 1's (2's) service sector is reduced (increased). It is not strict to have $dN_{i1}/dA_1 < 0$ and $dN_{i2}/dA_1 > 0$. This is guaranteed, for instance, in the case of $\beta\xi\Lambda/\gamma > a$ and $\beta\xi/\gamma > a$. For convenience of discussion, we assume

$$\min\left\{\frac{\beta\xi\Lambda}{\gamma}, \frac{\beta\xi}{\gamma}, \frac{\beta\eta}{\gamma}\right\} > a. \tag{8.3.6}$$

The number employed by region 1's (2's) industrial sector is reduced (increased). By (8.A.1.8) and (8.A.1.10), we get

$$\frac{1}{K_{s1}}\frac{dK_{s1}}{dA_1} = \frac{1}{N_{s1}}\frac{dN_{s1}}{dA_1} - \frac{\theta}{\Lambda}\frac{d\Lambda}{dA_1} - \frac{\eta}{\theta A_1} < 0,$$
$$\frac{1}{K_{i1}}\frac{dK_{i1}}{dA_1} = \frac{1}{N_{i1}}\frac{dN_{i1}}{dA_1} - \frac{\theta}{\Lambda}\frac{d\Lambda}{dA_1} - \frac{\eta}{\theta A_1} < 0,$$
$$\frac{1}{K_{s2}}\frac{dK_{s2}}{dA_1} = \frac{1}{N_{s2}}\frac{dN_{s2}}{dA_1} + \frac{\theta}{\Lambda}\frac{d\Lambda}{dA_1} - \frac{\eta}{\theta A_1},$$
$$\frac{1}{K_{i2}}\frac{dK_{i2}}{dA_1} = \frac{1}{N_{i2}}\frac{dN_{i2}}{dA_1} + \frac{\theta}{\Lambda}\frac{d\Lambda}{dA_1} - \frac{\eta}{\theta A_1}. \tag{8.3.7}$$

The capital stocks employed by region 1's two sectors are reduced and the capital stocks employed by region 2's two sectors may either be increased or decreased.

$$\frac{1}{F_{s1}}\frac{dF_{s1}}{dA_1} = \frac{1}{N_{s1}}\frac{dN_{s1}}{dA_1} + \frac{a_1\theta}{\Lambda}\frac{d\Lambda}{dA_1} - \frac{a_1\eta}{\theta A_1} < 0,$$
$$\frac{1}{F_{i1}}\frac{dF_{i1}}{dA_1} = \frac{1}{N_{i1}}\frac{dN_{i1}}{dA_1} + \frac{\alpha\theta}{\Lambda}\frac{d\Lambda}{dA_1} - \frac{\alpha\eta}{\theta A_1} < 0,$$
$$\frac{1}{F_{s2}}\frac{dF_{s2}}{dA_1} = \frac{1}{N_{s2}}\frac{dN_{s2}}{dA_1} + \frac{a_2\theta}{\Lambda}\frac{d\Lambda}{dA_1} - \frac{a_2\eta}{\theta A_1},$$
$$\frac{1}{F_{i2}}\frac{dF_{i2}}{dA_1} = \frac{1}{N_{i2}}\frac{dN_{i2}}{dA_1} + \frac{\alpha\theta}{\Lambda}\frac{d\Lambda}{dA_1} - \frac{\alpha\eta}{\theta A_1}. \tag{8.3.8}$$

Taking derivatives of $R_j = \eta\Omega N_j / L_j$ and (8.1.9) yields

$$\frac{1}{R_1}\frac{dR_1}{dA_1} = \frac{1}{\Omega}\frac{d\Omega}{dA_1} + \frac{1}{N_1}\frac{dN_1}{dA_1} < 0, \quad \frac{1}{R_2}\frac{dR_2}{dA_1} = \frac{1}{\Omega}\frac{d\Omega}{dA_2} + \frac{1}{N_2}\frac{dN_2}{dA_1},$$
$$\frac{1}{l_1}\frac{dl_1}{dA_1} = -\frac{1}{N_1}\frac{dN_1}{dA_1} > 0, \quad \frac{1}{l_2}\frac{dl_2}{dA_1} = -\frac{1}{N_2}\frac{dN_2}{dA_1} < 0,$$
$$\frac{1}{c_{js}}\frac{dc_{js}}{dA_1} = \frac{1}{\Omega}\frac{d\Omega}{dA_1} - (\alpha - a_j)\left(\frac{1}{\Lambda}\frac{d\Lambda}{dA_1} - \frac{1}{\eta A_1}\right),$$
$$\frac{1}{c_{ji}}\frac{dc_{ji}}{dA_1} = \frac{1}{\Omega}\frac{d\Omega}{dA_1} < 0. \tag{8.3.9}$$

We see that as region 1's amenity is improved, region 1's land rent tends to be reduced (in association with emigration); region 2's land rent may either be increased or reduced; the lot size per capita in region 1 (2) is increased (reduced) due to emigration (immigration); service consumption per capita in the two regions may either be increased or reduced; and the consumption levels per capita of industrial goods in the two regions are reduced. We conclude that in terms of the consumption levels per capita of goods, services, and housing an improvement in one region's amenity level benefits a household in relation to some variables and reduces its consumption levels in relation to other variables.

8.4 The Propensity to Hold Wealth and the Economic Structure

This section is concerned with the effects of changes in the propensity, λ, to hold wealth on the equilibrium structure of economic geography. By (8.2.4),

$$N_1 = \frac{\Lambda N}{1 + \Lambda}, \quad N_2 = \frac{N}{1 + \Lambda}$$

we have

$$\Phi' \frac{d\Lambda}{d\lambda} = - \frac{1/b_0^{\beta}\Lambda^{\beta\theta} - \delta_k}{\Lambda}\alpha < 0, \quad \frac{1}{N_1}\frac{dN_1}{d\lambda} = \frac{1}{\Lambda\Lambda_1}\frac{d\Lambda}{d\lambda} > 0,$$
$$\frac{1}{N_2}\frac{dN_2}{d\lambda} = - \frac{1}{\Lambda_1}\frac{d\Lambda}{d\lambda} < 0. \tag{8.4.1}$$

As the propensity to hold wealth is increased, some of region 2's population will migrate to region 1. Taking derivatives of $r = \alpha / b_0^{\beta}\Lambda^{\beta\theta}$, $w = \beta b_0^{\beta}\Lambda^{\alpha\theta}$, and (8.3.2), we get

$$\frac{1}{r}\frac{dr}{d\lambda} = - \frac{\beta\theta}{\Lambda}\frac{d\Lambda}{d\lambda} < 0, \quad \frac{1}{w}\frac{dw}{d\lambda} = \frac{\alpha\theta}{\Lambda}\frac{d\Lambda}{d\lambda} > 0,$$
$$\frac{1}{p_j}\frac{dp_j}{d\lambda} = (\alpha - a_j)\frac{\theta}{\Lambda}\frac{d\Lambda}{d\lambda}, \quad j = 1, 2. \tag{8.4.2}$$

As the propensity to hold wealth is increased, the rate of interest, r, is reduced, the wage rate, w, is increased and region j's services price, p_j, is increased (decreased) in the case that a_j is larger (smaller) than a. By $r = \alpha / b_0^{\beta} \Lambda^{\beta\theta}$, $w = \beta b_0^{\beta} \Lambda^{\alpha\theta}$, and (8.3.2), we have

$$\frac{1}{\Omega}\frac{d\Omega}{d\lambda} = \frac{K_0}{\Lambda}\frac{d\Lambda}{d\lambda}$$

$$\left\{\alpha\beta\theta\xi + \frac{(\alpha\theta b_2 + b_2 + \alpha\theta b_1 - b_1)\Lambda + \alpha\theta b_2 + \alpha\theta b_1\Lambda^2}{(1 + \Lambda)^2}\gamma\right\},$$

$$\frac{1}{K}\frac{dK}{d\lambda} = \frac{1}{\Omega}\frac{d\Omega}{d\lambda} + \frac{1}{\lambda} > 0 \tag{8.4.3}$$

where we require $\alpha\theta(b_1 + b_2) > b_1 - b_2$. The total capital stocks, K, is increased as the propensity to hold wealth is increased.

For simplicity, we require (8.3.7) is satisfied in the remainder of this section. By (8.A.1.11), (8.A.1.6), (8.A.1.8), $N_{ij} = N_j - N_{sj}$ and (8.A.1.10), we get

$$\frac{\Lambda_1}{N_{s1}}\frac{dN_{s1}}{d\lambda} = K_0\frac{\beta\xi + \gamma b_2}{\Lambda}\frac{d\Lambda}{d\lambda} > 0,$$

$$\frac{\Lambda_1}{N_{s2}}\frac{dN_{s1}}{d\lambda} = -K_0(\beta\xi + \gamma b_1)\frac{d\Lambda}{d\lambda} > 0,$$

$$\frac{n_1\Lambda_1^2}{N_{i1}}\frac{dN_{i1}}{d\lambda} = \{\beta\xi\Lambda^2 + 2\beta\xi\Lambda - \gamma(a_2 - a_1)\}\frac{\beta\xi + \gamma b_2}{\Lambda}\frac{d\Lambda}{d\lambda} > 0,$$

$$\frac{n_2\Lambda_1^2}{N_{i2}}\frac{dN_{i2}}{d\lambda} = -$$

$$\frac{\{2\beta\xi + (\beta\xi + a\gamma)\Lambda\}(\beta\xi + \gamma b_1)\Lambda + \beta^2\xi^2 + (\beta\xi - a\gamma)\gamma b_2}{\Lambda}\frac{d\Lambda}{d\lambda}. \tag{8.4.4}$$

The labor inputs, N_{s1} and N_{s2} (N_{i1} and N_{i2}), of region 1's (2's) two sectors are increased (reduced). By (8.A.1.8) and (8.A.1.10), we get

$$\frac{1}{K_{s1}}\frac{dK_{s1}}{d\lambda} = \frac{1}{N_{s1}}\frac{dN_{s1}}{d\lambda} + \frac{\theta}{\Lambda}\frac{d\Lambda}{d\lambda} > 0,$$
$$\frac{1}{K_{i1}}\frac{dK_{i1}}{d\lambda} = \frac{1}{N_{i1}}\frac{dN_{i1}}{d\lambda} + \frac{\theta}{\Lambda}\frac{d\Lambda}{d\lambda},$$
$$\frac{1}{K_{s2}}\frac{dK_{s2}}{d\lambda} =$$
$$\{(\beta\xi + \gamma b_1)(\theta - 1)\Lambda + \beta\theta\xi + \gamma\theta b_2\}\frac{K_0}{\Lambda_1\Lambda}\frac{d\Lambda}{d\lambda} > 0,$$
$$\frac{1}{K_{i2}}\frac{dK_{i2}}{d\lambda} = \frac{1}{N_{i2}}\frac{dN_{i2}}{d\lambda} + \frac{\theta}{\Lambda}\frac{d\Lambda}{d\lambda}. \tag{8.4.5}$$

The capital stocks employed by region 1's two sectors and by region 2's industrial sector are increased. It is not easy to explicitly determine the sign of $dK_{i2} / d\lambda$. Taking derivatives of (8.1.1) and (8.1.3) yields

$$\frac{1}{F_{s1}}\frac{dF_{s1}}{d\lambda} = \frac{1}{N_{s1}}\frac{dN_{s1}}{d\lambda} + \frac{a_1\theta}{\Lambda}\frac{d\Lambda}{d\lambda} > 0,$$
$$\frac{1}{F_{i1}}\frac{dF_{i1}}{d\lambda} = \frac{1}{N_{i1}}\frac{dN_{i1}}{d\lambda} + \frac{\alpha\theta}{\Lambda}\frac{d\Lambda}{d\lambda} > 0,$$
$$\frac{1}{F_{s2}}\frac{dF_{s2}}{d\lambda} = \frac{1}{N_{s2}}\frac{dN_{s2}}{d\lambda} + \frac{a_2\theta}{\Lambda}\frac{d\Lambda}{d\lambda},$$
$$\frac{1}{F_{i2}}\frac{dF_{i2}}{d\lambda} = \frac{1}{N_{i2}}\frac{dN_{i2}}{d\lambda} + \frac{\alpha\theta}{\Lambda}\frac{d\Lambda}{d\lambda}. \tag{8.4.6}$$

We conclude that the output levels of region 1's two sectors are increased; the output levels of region 2's two sectors may be either increased or reduced. Taking derivatives of $R_j = \eta\Omega N_j / L_j$ and (8.1.9) with respect to λ yields

$$\frac{1}{R_1}\frac{dR_1}{d\lambda} = \frac{1}{\Omega}\frac{d\Omega}{d\lambda} + \frac{1}{\Lambda\Lambda_1}\frac{d\Lambda}{d\lambda} > 0, \quad \frac{1}{R_2}\frac{dR_2}{d\lambda} = \frac{1}{\Omega}\frac{d\Omega}{d\lambda} - \frac{1}{\Lambda_1}\frac{d\Lambda}{d\lambda},$$
$$\frac{1}{l_1}\frac{dl_1}{d\lambda} = -\frac{1}{N_1}\frac{dN_1}{d\lambda} < 0, \quad \frac{1}{l_2}\frac{dl_2}{d\lambda} = -\frac{1}{N_2}\frac{dN_2}{d\lambda} > 0,$$

$$\frac{1}{c_{js}}\frac{dc_{js}}{d\lambda} = \frac{1}{\Omega}\frac{d\Omega}{d\lambda} - \left(\alpha - a_j\right)\frac{\theta}{\Lambda}\frac{d\Lambda}{d\lambda},$$

$$\frac{1}{c_{ji}}\frac{dc_{ji}}{d\lambda} = \frac{1}{\Omega}\frac{d\Omega}{d\lambda} > 0. \tag{8.4.7}$$

We see that as the propensity to hold wealth is increased, region 1's land rent tends to be increased (in the association with immigration), region 2's land rent may be either increased or reduced, the lot size per capita in region 1 (2) is reduced (increased) due to the immigration (emigration), service consumption per capita in the two regions may be either increased or reduced, the consumption levels per capita of industrial goods in the two regions are increased.

8.5 On Regional Growth and Structural Change

We propounded a compact dynamic two-region economic model with endogenous capital accumulation. The model recounts the interaction between capital accumulation, the regional distribution of capital and labor, the division of labor, the region distribution of capital, land rents, regional service prices and commodity prices over time and space. We guaranteed the existence of equilibria with the given preference structure, the regional amenities, technologies, and resources (of land and labor) and analyzed how the differences in regional resources such as amenities and the preference structure of the population may affect the equilibrium structure of economic geography.

It is possible to extend the model to economies with returns to scale effect; for instance, by introducing infrastructures, population or knowledge dynamics into the system. It is well known that an economic system with returns to scale variables such as knowledge and/or population tends to have multiple equilibria. We may also further disaggregate economic structures and consider geographical structures within each region. Since we are concerned with a highly aggregated model, the omission of transportation costs may not essentially affect our results if the transportation cost is not large in comparison to the total industrial product. If the transportation cost is high, we may introduce transportation cost factors by assuming that the cost for transporting goods from one region to another region is linearly proportional to the amount of goods transported (i.e., Samuelson's iceberg assumption). We may include national and/or local tax systems to supply, for instance, public goods. It is expected that the difference between the nature of the public goods and the tax rates in the regions may strongly affect economic geography. For instance, if we assume that the regional government can tax its region's households and producers and that the tax income is used for improving amenity, then the regional amenity will become an endogenous variable. It can be intuitively seen that it is difficult to judge the impact of such a regional policy because of its complicated impact on the whole system. We will discuss these issues elsewhere.

Appendix

A.8.1 Proving Lemma 8.2.1

Substituting (8.1.9) into (8.1.7) and then using $C_{ji} = c_{ji} N_j$ and $C_{js} = c_{js} N_j$, we obtain

$$\frac{p_1}{p_2} = A \left(\frac{R_2}{R_1} \right)^{\eta / \gamma} \tag{8.A.1.1}$$

where $A \equiv (A_1 / A_2)^{1/\gamma}$. By (8.1.2) and (8.1.4), we get

$$\frac{K_{s1}}{N_{s1}} = \frac{a_1 b_2}{a_2 b_1} \frac{K_{s2}}{N_{s2}}. \tag{8.A.1.2}$$

By $l_j N_j = L_j$ in (8.1.16) and $R_j N_j = \eta \Omega$ in (8.1.9), we obtain

$$\frac{N_1}{N_2} = \frac{R_1 L_1}{R_2 L_2}. \tag{8.A.1.3}$$

By $p_j c_{js} = \gamma \Omega$ in (8.1.9), $C_{js} = c_{js} N_j$, and (8.1.14), we get

$$\gamma N_j \Omega = p_j F_{sj}. \tag{8.A.1.4}$$

By (8.A.1.3) and (8.A.1.4), we get

$$\frac{p_1 F_{s1}}{p_2 F_{s2}} = \frac{R_1 L_1}{R_2 L_2}.$$

Substituting (8.A.1.1) into this equation yields

$$\left(\frac{p_1}{p_2}\right)^{1+\gamma/\eta}\frac{F_{s1}}{F_{s2}} = A^{\gamma/\eta}\frac{L_1}{L_2}. \tag{8.A.1.5}$$

By $w = b_j p_j F_{sj} / N_{sj}$ in (8.1.4) and (8.A.1.4), we get

$$\frac{N_{s1}}{N_{s2}} = \Lambda\frac{b_1}{b_2} \tag{8.A.1.6}$$

where $\Lambda \equiv N_1 / N_2$. By (8.1.4), we have

$$\frac{p_1}{p_2} = \frac{b_2 N_{s1} F_{s2}}{b_1 N_{s2} F_{s1}}.$$

Substituting this equation, (8.1.3) and (8.A.1.6) into (8.A.1.5), we get

$$\left(\frac{N_{s1}}{K_{s1}}\right)^{\gamma\alpha_1/\eta}\left(\frac{K_{s2}}{N_{s2}}\right)^{\gamma\alpha_2/\eta} = \left(\frac{Ab_1}{b_2}\right)^{\gamma/\eta}\frac{\Lambda L_1}{L_2}. \tag{8.A.1.7}$$

By (8.A.1.2) and (8.A.1.7), we solve K_{sj} / N_{sj} as functions of Λ as

$$\frac{K_{s1}}{N_{s1}} = \frac{a_1 b_2}{a_2 b_1} b\Lambda^\theta, \quad \frac{K_{s2}}{N_{s2}} = b\Lambda^\theta \tag{8.A.1.8}$$

where

$$\theta = \frac{\eta}{a\gamma} > 0, \quad b = \left(\frac{a_1 b_2}{a_2 b_1}\right)^{\gamma\theta\alpha_1/\eta}\left(\frac{Ab_1}{b_2}\right)^{\gamma\theta/\eta}\frac{L_1}{L_2}.$$

By (8.1.2) and (8.1.4), we have

$$\frac{K_{ij}}{N_{ij}} = \frac{a_2 b}{a b_2}\frac{K_{s2}}{N_{s2}}.$$

From this equation and (8.A.1.8), we get

$$\frac{K_{i1}}{N_{i1}} = \frac{K_{i2}}{N_{i2}} = b_0 \Lambda^{\theta} \tag{8.A.1.9}$$

where $b_0 \equiv a_2 b\beta / \alpha b_2$. By $N / N_1 = L_1 / N$, (8.A.1.4) and

$$p_1 F_{s1} = \frac{wF_{s1}}{b_1}$$

we have

$$\gamma \Omega N = \left(1 + \frac{1}{\Lambda}\right) \frac{wF_{s1}}{b_1}. \tag{8.A.1.10}$$

Substituting $F_{ij} = wN_{ij} / \beta$, $C_{ij} = \xi \Omega N_j$, and $s_j = \lambda \Omega - (1 - \delta_k) k$ into (8.1.17) yields

$$N_{i1} + N_{i2} = \{(\xi + \lambda)\Omega - (1 - \delta_k) k\} \frac{\beta}{w}.$$

By this equation, $N_{i1} - N_{i2} = N - N_{s1} - N_{s2}$ and (8.A.1.6), we have

$$N_{s1} = \frac{N}{1 + b_2 / b_1 \Lambda} - \frac{(\xi + \lambda)\Omega N - (1 - \delta_k) K}{1 + b_2 / b_1 \Lambda} \frac{\beta}{w}. \tag{8.A.1.11}$$

By $R_j l_j = \eta \Omega$, (8.1.16) and (8.1.5), we have

$$R_0 = \eta \Omega.$$

By $R_0 = \eta \Omega$, $Y = yN$ and (8.1.6), we have

$$(1 - \eta \rho) Y = \eta \delta K + wN + rK.$$

Substituting the above equation and $\Omega N = \rho Y + \delta K$, we have

$$\Omega N = \frac{\rho w N + \rho r K + \delta K}{1 - \eta \rho}. \tag{8.A.1.12}$$

By the above equation and (8.A.1.11) into (8.A.1.10) we have

$$H(\Lambda) = \frac{\beta\xi + \beta\lambda + (b_1\Lambda + b_2)\gamma / \Lambda_1}{b_0^\beta \Lambda^{\beta\theta}}\alpha K$$

$$- \frac{\beta\lambda - (b_1\Lambda + b_2)/\Lambda_1}{\rho}\delta\gamma K -$$

$$\left\{\alpha\xi + \alpha\lambda + (a_1\Lambda + a_2)\frac{\gamma}{\Lambda_1}\right\}\beta b_0^\beta \Lambda^{\beta\theta} N = 0 \qquad (8.A.1.13)$$

where we use $r = \alpha / b_0^\beta \Lambda^{\beta\theta}$ and $w = \beta b_0^\beta \Lambda^{\alpha\theta}$, which are obtained by (8.1.2) and (8.A.1.9). The equation has only two variables, K and Λ. We have

$$H(0) > 0, \quad H(\infty) < 0, \quad \forall\ K > 0 .$$

This guarantees that the equation, $H(\Lambda) = 0$, has at least one positive solution for any given positive K. Taking derivatives of (8.A.1.12) with respect to Λ, we have

$$\Lambda_1^2 \Phi^*(\Lambda) = \frac{\alpha\delta\gamma K}{\rho} - \beta N \Lambda_1 b_0^\alpha \Lambda^{\beta\theta - 1} - \frac{\alpha\Lambda_2 K}{b_0^\beta \Lambda^{\beta\theta+1}} \qquad (8.A.1.14)$$

where $\Phi^* \equiv dH / d\Lambda$ and

$$\Lambda_1 \equiv \alpha\theta\gamma\Lambda(a_1\Lambda + a_2) - \alpha\gamma\Lambda + \alpha^2\theta\Lambda^2 ,$$

$$\Lambda_2 \equiv \beta^2\theta\Lambda_1^2(\xi + \lambda) + \beta\gamma\theta\Lambda_1(b_1\Lambda + b_2) - \gamma\Lambda(b_1 - b_2).$$

It is not strict to require $\Lambda_1 > 0$ and $\Lambda_2 > 0$. We thus conclude that for any given positive $K(t)$ at any point of time, (8.A.1.13) has at least one solution, $\Lambda = \Lambda(K)$. When a is sufficiently small, the solution is unique. For any $\Lambda > 0$, all the other variables are uniquely determined as functions of Λ at any point of time by the following procedure: $r = \alpha / b_0^\beta \Lambda^{\beta\theta}$ and $w = \beta b_0^\beta \Lambda^{\alpha\theta}$ $\rightarrow$ Ω by (8.A.1.12) $\rightarrow$ $Y = (\Omega N - \delta K)/\rho$ and $y = Y / N$ $\rightarrow$ N_{s1} by (8.A.1.11) $\rightarrow$ N_{s2} by (8.A.1.6) $\rightarrow$ $N_1 = N\Lambda/(1 + \Lambda)$ and $N_2 = N/(1 + \Lambda)$ $\rightarrow$ $N_{ij} = N_j - N_{sj}$, $j = 1, 2$ $\rightarrow$ K_{s1} by (8.A.1.8) $\rightarrow$ K_{ij} by (8.A.1.9) $\rightarrow$ F_{ij} by (8.1.16) $\rightarrow$ F_{sj} by (8.1.3) $\rightarrow$ p_j by (8.1.4) and

(8.A.1.8) → $R_j = \eta\Omega N_j / L_j$ by (8.1.9) and (8.1.16) → l_j, c_{js}, c_{ji}, and s_j by (8.1.9) → C_{ji} and C_{js} by (8.1.13) → U_j by (8.1.7). If (8.A.1.13) has a unique solution, the functions are unique. We proved Lemma 8.2.1.

9 A Two-region Growth Model with Capital and Knowledge

The previous chapter developed models with knowledge and human capital fixed. This chapter treats knowledge as endogenous variables of regional economic evolution. Knowledge creation and utilization is the driving force of modern economic development. A main issue in economics is concerned with dynamic interactions between economic growth, knowledge creation and utilization. The complexity of economic dynamics with endogenous knowledge has recently caused great attention among economists. Although many growth models with endogenous knowledge have been proposed, most of those models are either concerned with national growth or international trades. There are only a few dynamic models which are constructed to address issues related to regional development with endogenous knowledge.

The literature on economic development has recently been centered on identifying different aspects of dynamic interactions between growth and knowledge accumulation. One of the first seminal attempts to render technical progress endogenous in growth models was initiated by Arrow (1962) emphasizing one aspect of knowledge accumulation - learning by doing. Uzawa (1965) introduced a sector specifying in creating knowledge into growth theory (see also, Zhang, 1990a, 1991a, 1999, Aghion and Howitt, 1992, 1998). The knowledge sector utilizes labor and the existing stock of knowledge to produce new knowledge, which enhances productivity of the production sector. Another approach was taken by, for instance, Kennedy (1964), Weisäcker (1966), Drandakis and Phelps (1966), and Samuelson (1965), who took account of the assumption of "inducement through the factor prices". Lucas (1988) emphasized the incentive effects of policy on investment in human capital. There are many other studies on endogenous technical progresses (e.g., Robson, 1980, Sato and Tsutsui, 1984, Nelson and Winter, 1982, Dosi, Pavitt and Soete, 1990, Andersson, 1986, Andersson, Anderstig, and Hårsman, 1990, Johansson and Karlsson, 1990). In Romer (1986), knowledge is taken as an input in the production function and competitive equilibrium is rendered consistent with increasing aggregate returns owing to externalities. It is assumed that knowledge displays increasing marginal productivity but new knowledge is produced by investment in research, which exhibits diminishing returns. Various other issues related to innovation, diffusion of technology and behavior of economic agents under various institutions have been discussed in the literature. There are also many other models emphasizing different aspects, such as education, trade, R&D policies, entrepreneurship, division of labor, learning through trading, brain drain, economic geography, of dynamic interactions among economic structure, development and knowledge (e.g., Dollar, 1986, Stokey, 1991, Nijkamp et al., 1998, Barro and Sal-i-Martin, 1995). It may be argued that there is a theoretical limitation in those works. Capital and knowledge have not been integrated into a compact theoretical framework with economic structures and free markets. For instance, in the works by Grossman and Helpman (1991) physical capital is almost totally neglected. This chapter proposes a compact framework to explain urban dynamics with endogenous urban pattern and capital and knowledge accumulation.

The objective of this chapter is to study relationships between regional growth and regional trade patterns in the context of perfectly competitive institution. The national economy consists of two regions. Each country produces goods and makes contribution to knowledge growth. Each region's production is similar to the standard one-sector growth model. Knowledge accumulation is through learning by doing (e.g., Arrow, 1962). The regional trade aspects of our model is similar to the international macroeconomic one-sector growth model with perfect capital mobility. It is well known that many efforts have been made to examine the impact of savings, technology and various policies on the trade patterns within the framework of dynamic models with capital accumulation (e.g., Oniki and Uzawa, 1965, Buiter, 1981, Findlay, 1984). Our modeling framework is based on the dynamic one-sector and two-country framework (see also, e.g., Zhang, 1990b, Wang, 1990, Ikeda and Ono, 1992). We are interested in how difference in knowledge utilization efficiency between two regions may affect pattern of regional trade and national economic development. The main differences between the models just mentioned and this chapter are related to labor mobility. This chapter assumes free mobility of labor force between regions.

This chapter is organized as follows. Section 9.1 defines the basic model. Section 9.2 provides the conditions for existence of equilibria of the dynamic system. Section 9.3 examines the effects of changes in each region's amenity level, the population and each region 1's knowledge accumulation efficiency on the equilibrium structure of the regional economy. Section 9.4 concludes the chapter. In Appendixes A.1 and A.2, we study the impact of changes in the population and each region's knowledge accumulation efficiency on the equilibrium structure. This chapter is based on Zhang (1997b).

9.1 Regional Growth with Endogenous Knowledge

We consider an economic system consisting of two regions, indexed by 1 and 2, respectively. Each region has a single production sector. Production is carried out by combining capital, labor and knowledge. It is assumed that the two regions' product is qualitatively homogenous and is either consumed or invested. The assumption of a single product used either for consumption or production is similar to the neo-classical dynamic one-sector trade models with capital (e.g., Frenkel and Razin, 1987, Ruffin, 1979). We assume a homogenous population. A person is free to choose his residential location. We assume that any person chooses the same region where he works and lives. It is assumed that climates and environment are homogenous within each region, but they may be different between the two regions. We select commodity to serve as numeraire. The amenity levels are assumed to regionally fixed. Since people's locational choice is affected by regional environmental conditions and the temporary equilibrium condition for labor movement is that people achieve the same level of utility in two regions, we see that wage rates between the two regions may not be equal.

To describe the model, we introduce

N — the given population of the economy;
$K(t)$ — the total capital stocks of the economy at time t;
$Z(t)$ — the level of knowledge stock at time t;
$F_j(t)$ — the output levels of region j 's production sector;
$K_j(t)$ and $N_j(t)$ — the levels of capital stocks and labor force employed by region j 's production sector;
$c_j(t)$ and $s_j(t)$ — the consumption level of and level of savings per capita in region j;
$y_j(t)$ — the net income per capita in region j;
$r(t)$ — the rate of interest; and
$w_j(t)$ — region j 's wage rate.

We now describe the model.

Regions' production
It is assumed that production functions of the two regions are given as follows

$$F_j(t) = Z^{m_j} K_j^{\alpha} N_j^{\beta}, \quad m_j \geq 0, \quad \alpha, \beta > 0, \quad \alpha + \beta = 1, \quad j = 1, 2 \tag{9.1.1}$$

in which m_j are region j 's knowledge utilization efficiency parameters. They measure how effectively each region utilizes the knowledge reservoir of the economy. The marginal conditions are given by

$$r = \frac{\alpha F_j}{K_j}, \quad w_j = \frac{\beta F_j}{N_j}. \tag{9.1.2}$$

The rate of interest is identical in the whole economy and the wage rates, w_j, may be different between the two regions.

Behavior of households

If we denote $k_j(t)$ the level of capital stocks per capita in region j, the interest payment per capita is given by $r(t)k_j(t)$. The net income per capita, $y_j(t)$ in region j consists of the wage income, $w_j(t)$, and interest payment, $r(t)k_j(t)$, i.e.

$$y_j(t) = w_j(t) + r(t)k_j(t), \quad j = 1, 2. \tag{9.1.3}$$

It is assumed that a typical person's utility level, $U_j(t)$, in region j is dependent on the person's consumption level, $c_j(t)$, of community, and the level of the person's wealth, $k_j(t) + s_j(t) - \delta_k k_j(t)$, where δ_k is the fixed depreciation rate of capital. The utility functions are specified as follows

$$U_j(t) = A_j c_j^{\xi_j} \left(k_j + s_j - \delta_k k_j\right)^{\lambda_j}, \quad \xi_j, \lambda_j > 0, \quad \xi_j + \lambda_j = 1, \quad j = 1, 2 \tag{9.1.4}$$

in which ξ_j and λ_j are respectively region j 's propensities to consume commodity and to hold wealth.

Since a family consists of several members and each member has his/her own utility function, family's behavior should be analyzed as the result of all members' rational decisions. The "collective" utility function should be analyzed within a framework which explicitly takes accounts of interactions within family's members (e.g., Becker, 1976, Chiappori, 1988). As far as regional economic dynamics are concerned, an explicit consideration of differences in locational preferences between family members is important. For instance, the husband and the wife may like to live in different regions. In such a situation, it is necessary to decompose the family structure in order to analyze how each member's preference and decision power may affect the family's location. Here, for simplicity of analysis, we neglect issues related to multiple members in locational decision.

In this chapter, we assume that the two regions may have different propensities to consume goods and to hold wealth. People's propensities to, for instance, to eat and to hold wealth may be different between regions with hot and cold weather. People's preference for cultural goods may be also spatially different. In (9.1.4), A_j is called region j 's amenity level. In this chapter, amenity levels are considered as locational parameters or slow changing variables, such as infrastructures, regional cultures and climates, which may affect the attractiveness of a location. It is not difficult to propose some reasonable assumptions to make amenities as endogenous variables represented in the form of

$$\frac{dA_j}{dt} = D_j(F, N, A), \quad j = 1, 2$$

where D_j are proper functions, F , N^* , and A are respectively the vectors of regional output levels, the population distribution and the amenity levels. For instance, it may be reasonable to assume that an increase in F_j reduces region j 's amenity level due to pollution. If we specify D_j properly, we can take account of endogenous amenity within our analytical framework.

A household's income is distributed between consumption and saving. The budget constraints are given by

$$c_j + s_j = y_j, \quad j = 1, 2. \tag{9.1.5}$$

A typical person maximizes his utility subject to the budget constraint, (9.1.5). The optimal problem has the following unique solution

$$c_j = \xi_j \Omega_j, \quad s_j = \lambda_j \Omega_j - \delta k_j \tag{9.1.6}$$

where

$$\Omega_j \equiv y_j + dk_j, \quad d \equiv 1 - \delta_k .$$

According to the definitions of k_j and s_j, the capital accumulation of a typical person in region j is given by

$$\frac{dk_j}{dt} = s_j - \delta k_j .$$

Substituting s_j in (9.1.6) into these two equations yields

$$\frac{dk_j}{dt} = \lambda_j \Omega_j - k_j . \tag{9.1.7}$$

As people are freely mobile between the two regions, the utility level of people should be equal, irrespective of in which region they live, i.e.,

$$U_1(t) = U_2(t) . \tag{9.1.8}$$

This equation is the temporary equilibrium condition for interregional labor force markets.

Knowledge accumulation
For simplicity of discussion, we only take account of learning by doing in formulating knowledge accumulation. We propose the following possible dynamics of knowledge

$$\frac{dZ}{dt} = \frac{\tau_1 F_1}{Z^{\varepsilon_1}} + \frac{\tau_2 F_2}{Z^{\varepsilon_2}} - \delta_z Z \tag{9.1.9}$$

in which τ_j (≥ 0), ε_j, and δ_z (≥ 0) are parameters.

Here, we interpret $\tau_j F_j / Z^{\varepsilon_j}$ as region j 's contribution to knowledge accumulation through learning by doing. We assume that the contribution to the knowledge creation of region j 's labor force is positively and linearly related to the region production scale, F_j. The term, $1/Z^{\varepsilon_j}$, implies that region j 's knowledge accumulation exhibits return to scale effects. The parameters, ε_j, measure return to scale effects of knowledge in knowledge accumulation by region j 's labor force. We say that the contribution to knowledge growth of region j exhibits increasing (decreasing) scale effects when $\varepsilon_j < (>) \ 0$. We interpret τ_j as a measurement of knowledge accumulation efficiency. It should be noted that it is conceptually not difficult to introduce other ways, such as research institutions and education, of accumulating knowledge.

Balance of demand and supply
By the definitions of K, k_j and N_j, we have

$$K = k_1 N_1 + k_2 N_2 . \tag{9.1.10}$$

The above equation tells that the total capital stocks of the economy is equal to the sum of the capital stocks owned by the two regions. The assumption that labor force and capital stocks are fully employed is represented by

$$N_1 + N_2 = N, \quad K_1 + K_2 = K . \tag{9.1.11}$$

The product is either consumed or saved. The balance of demand for and supply of industrial commodity is given by

$$(c_1 + s_1)N_1 + (c_2 + s_2)N_2 = F_1 + F_2 . \tag{9.1.12}$$

This equation means that the sum of the two regions' consumption and savings is equal to the total product of the economy.

We completed the construction of the basic model. The model is built with some strict assumptions. It is structurally quite general in the sense that some well-established models can be considered as its special case. For instance, if we neglect any difference in production and preference and omit endogenous technology, then the model is identical with the Solow-Swan one-sector neoclassical growth model. If we neglect any difference in production and preference, then the model is similar to the Arrow's learning by doing model. If we don't allow labor mobility (i.e., fix labor force of each region and allow goods to be freely mobile), then the model is similar to the neoclassical trade model with endogenous population. The system has 21 variables, N_j, K_j, F_j, c_j, s_j, k_j, y_j, U_j, w_j ($j = 1, 2$), K, Z, and r. We now examine conditions for the existence of equilibria of the dynamic system.

9.2 Knowledge and the Equilibrium Structure

This section is concerned with conditions for the existence of equilibria of the dynamic system. It should be mentioned that since the system consists of so many variables and equations, it is not easy to interpret some equations obtained below. We are thus mainly concerned with solving the equilibrium issue.

By (9.1.9) and (9.1.7), at equilibrium we have

$$\frac{\tau_1 F_1}{Z^{\varepsilon_1}} + \frac{\tau_2 F_2}{Z^{\varepsilon_2}} = \delta_z Z , \quad \lambda_j \Omega_j = k_j, \quad j = 1, 2 . \tag{9.2.1}$$

Substituting $\lambda_j \Omega_j = k_j$ in (9.2.1) into (9.1.6) yields

$$c_j = \frac{\xi_j k_j}{\lambda_j}, \quad s_j = \delta_k k_j. \tag{9.2.2}$$

These equations tell that at a steady state the level of consumption per capita in region j is proportional to the capital stocks per capita. Substituting (9.2.2) into (9.1.4) and then using (9.1.8), we get

$$\frac{k_1}{k_2} = A \equiv \frac{A_2}{A_1}\left(\frac{\xi_2}{\lambda_2}\right)^{\xi_2}\left(\frac{\xi_1}{\lambda_1}\right)^{-\xi_1}. \tag{9.2.3}$$

We conclude that if the system has a steady state, then the ratio of the capital stocks per capita between region 1 and 2 is proportional to the ratio of the amenity levels between region 2 and 1. This implies that if one region's amenity is improved in comparison to the other region, then the capital stocks per capita in this region should be reduced in comparison to in the other region in order to maintain the labor markets at equilibrium. From (9.2.3), we see also that the ratio, k_1 / k_2, is nonlinearly related to the preferences parameters, λ_j and ξ_j.

By (9.2.2) and (9.1.12), we get

$$\delta_1 k_1 N_1 + \delta_2 k_2 N_2 = F_1 + F_2 \tag{9.2.4}$$

where

$$\delta_j \equiv \frac{\xi_j}{\lambda_j} + \delta_k.$$

By $\lambda_j \Omega_j = k_j$ in (9.2.1), (9.1.3) and the definitions of Ω_j, we get

$$w_j = (\delta_j - r)k_j, \quad j = 1, 2. \tag{9.2.5}$$

By (9.1.1) and (9.1.2), we get

$$\Lambda \equiv \frac{K_1}{N_1 Z^{m_1/\beta}} = \frac{K_2}{N_2 Z^{m_2/\beta}}, \quad r = \frac{\alpha}{\Lambda^\beta}, \quad w_j = \beta Z^{m_j/\beta} \Lambda^\alpha. \tag{9.2.6}$$

Substituting $F_j = rK_j / \alpha = K_j / \Lambda^\beta$ into (9.2.4) yields

$$\delta_1 k_1 N_1 + \delta_2 k_2 N_2 = \frac{K}{\Lambda^\beta} \tag{9.2.7}$$

where we use $K = K_1 + K_2$. Substituting (9.1.10) and (9.2.3) into (9.2.7) yields

$$\Gamma = \frac{\delta_2 \Lambda^\beta - 1}{(1 - \delta_1 \Lambda^\beta) A} \tag{9.2.8}$$

where $\Gamma \equiv N_1 / N_2$. By $N_1 + N_2 = N$ and $\Gamma = N_1 / N_2$, we have

$$N_1 = \frac{\Gamma N}{1 + \Gamma}, \quad N_2 = \frac{N}{1 + \Gamma}. \tag{9.2.9}$$

Dividing the first equation in (9.2.5) by the second one and then using (9.2.3) and (9.2.6), we obtain

$$Z^m = \frac{\delta_1 \Lambda^\beta - \alpha}{\delta_2 \Lambda^\beta - \alpha} \tag{9.2.10}$$

where

$$m \equiv \frac{m_1 - m_2}{\beta}.$$

In the case of $m = 0$, we solve Λ and Γ respectively by (9.2.10) and (9.2.8). If without special mention, $m \neq 0$ holds below.

For simplicity of discussion, we require: $\delta_2 / \alpha > \delta_1 > \delta_2$. We interpret this requirement later on. By (9.2.5) and $r = \alpha / \Lambda^b$ in (9.2.6), we see that $w_j > 0$ are guaranteed if

$$\Lambda^\beta > \max\left\{\frac{\alpha}{\delta_j}, \ j = 1, 2\right\} = \frac{\alpha}{\delta_2}.$$

By (9.2.10) we see that this condition also guarantees $Z^m > 0$. By (9.2.8) $\Gamma > 0$ is guaranteed if $1/\delta_2 > \Gamma > 1/\delta_1$. It is thus sufficient for us to require

$$\frac{1}{\delta_2} > \Gamma > \max\left\{\frac{1}{\delta_1}, \frac{\alpha}{\delta_2}\right\} = \frac{1}{\delta_1}. \tag{9.2.11}$$

By the first equation in (9.2.6) and $F_j = K_j / \Lambda^\beta$, we have

$$K_j = \Lambda N_j Z^{m_j/\beta}, \quad F_j = \Lambda^\alpha N_j Z^{m_j/\beta}. \tag{9.2.12}$$

Substituting F_j in (9.2.12) into the first equation in (9.2.1), we get

$$\Phi(\Lambda) \equiv \Phi_1(\Lambda) + \Phi_2(\Lambda) - d = 0, \quad \frac{1}{\delta_2} > \Lambda^\beta > \frac{1}{\delta_1} \tag{9.2.13}$$

where we use (9.2.8) and

$$\Phi_1(\Lambda) \equiv \frac{\tau_1 N\Gamma Z^{x_1}\Lambda^\alpha}{1+\Gamma}, \quad \Phi_2(\Lambda) \equiv \frac{\tau_2 N Z^{x_2}\Lambda^\alpha}{1+\Gamma},$$

$$x_j \equiv \frac{m_j}{\beta} - \varepsilon_j - 1, \quad j = 1, 2.$$

In (9.2.13), Γ and Z are functions of Λ respectively defined by (9.2.8) and (9.2.10). By (9.2.13), (9.2.8), and (9.2.10), we conclude that the function, $\Phi(\Lambda)$, has the following properties

$$\Phi(\frac{1}{\delta_2^{1/\beta}}) = \frac{\{(\delta_1/\delta_2 - \alpha)A/\beta\}^{x_2/\beta}}{\delta_2^{\alpha/\beta}}\tau_2 N - d,$$

$$\Phi(\frac{1}{\delta_1^{1/\beta}}) = \frac{\{\beta A/(\delta_2/\delta_1 - \alpha)\}^{x_1/\beta}}{\delta_1^{\alpha/\beta}}\tau_1 N - d,$$

$$\frac{d\Phi}{d\Lambda} = \frac{x_1\Phi_1 + x_2\Phi_2}{Z}\frac{dZ}{d\Lambda} + \frac{\Phi_1/\Gamma - \Phi_2}{1+\Gamma}\frac{d\Gamma}{d\Lambda} + \alpha\frac{\Phi_1 + \Phi_2}{\Lambda^\beta} \tag{9.2.14}$$

in which

$$m\frac{dZ}{d\Lambda}=\frac{\delta_2-\delta_1}{\left(\delta_2\Lambda^\beta-\alpha\right)\left(\delta_1\Lambda^\beta-\alpha\right)}\frac{\alpha\beta Z}{\Lambda^\alpha}<0,$$

$$\frac{\Lambda^\alpha}{\beta\Gamma}\frac{d\Gamma}{d\Lambda}=\frac{\delta_2}{\delta_2\Lambda^\beta-\alpha}+\frac{\delta_1}{\delta_1\Lambda^\beta-\alpha}>0.$$

As it is difficult to explicitly judge the signs of $\Phi(1/\delta_1^{1/\beta})$, $\Phi(1/\delta_2^{1/\beta})$ and $d\Phi/d\Lambda$, we see that it is difficult to judge whether or not the equation, $\Phi(\Lambda)=0$, has solutions for $1/\delta_2>\Lambda^\beta>1/\delta_1$. We may solve the equation properly specifying values of some parameters. For instance, if $\Phi(1/\delta_2^{1/\beta})>0$ (which is guaranteed, for instance, if τ_2 is large), $\Phi(1/\delta_2^{1/\beta})<0$ (which is guaranteed, for instance, if τ_1 is small), and $d\Phi/d\Lambda>0$ (which is guaranteed, for instance, if $m>0$, $x_1<0$ and $x_2<0$), then the dynamic system has a unique equilibrium. Summarizing the above discussion, we obtain the following proposition.

Proposition 9.2.1.
We assume $m\geq 0$ and $\delta_2/\alpha>\delta_1>\delta_2$. If (9.2.13) has solutions for $1/\delta_2>\Lambda^\beta>1/\delta_1$, for any given solution, the dynamic system has a unique equilibrium. The number of economic equilibria is equal to the number of solutions of (9.2.13).

Since it is difficult to interpret the conditions that (9.2.13) has a meaningful solution, for illustration we now examine the case of $m=0$. We still require: $\delta_2/\alpha>\delta_1>\delta_2$. By (9.2.10), we solve Λ^β as follows

$$\Lambda^\beta=\frac{\alpha(A-1)}{\delta_1 A-\delta_2}. \tag{9.2.15}$$

The unique solution, Λ^β, is positive and satisfies (9.2.11) if

$$\frac{\delta_2/\delta_1-\alpha}{\beta}<A<\min\left\{\frac{\delta_2}{\delta_1},\frac{\beta}{\delta_2/\delta_1-\alpha}\right\}<1. \tag{9.2.16}$$

Under (9.2.16), we solve Λ by (9.2.15). By (9.2.8), we solve Γ. It should be noted that the values of Γ and Λ are independent of Z. Similarly to (9.2.13), we get

$$\Phi_0(Z) \equiv \Phi_{01}(Z) + \Phi_{02}(Z) - d = 0, \quad \infty > Z > 0 \tag{9.2.17}$$

where

$$\Phi_{01}(Z) \equiv \frac{\tau_1 N \Gamma Z^{x_1} \Lambda^{\alpha}}{1+\Gamma}, \quad \Phi_{02}(Z) \equiv \frac{\tau_2 N Z^{x_2} \Lambda^{\alpha}}{1+\Gamma}.$$

It should be noted that in (9.2.17) Λ and Γ are treated as parameters. The equation includes a single variable, Z.

We obtain that $\Phi_0(Z)$ has the following properties: 1) in the case of $x_j < 0$, we have

$$\Phi_0(Z) > 0, \quad \Phi_0(\infty) < 0, \quad \frac{d\Phi_0}{dZ} < 0, \quad \infty > Z > 0$$

2) in the case of $x_j > 0$, we have

$$\Phi_0(Z) < 0, \quad \Phi_0(\infty) > 0, \quad \frac{d\Phi_0}{dZ} > 0, \quad \infty > Z > 0$$

3) in the case of $x_1 > 0$ and $x_2 < 0$ ($x_1 < 0$ and $x_2 > 0$), we have

$$\Phi_0(Z) > 0, \quad \Phi_0(\infty) > 0,$$
$$\frac{d\Phi_0}{dZ} = 0 \; \textit{has a unique solution}, \quad \infty > Z > 0.$$

From these properties of $\Phi_0(Z)$, we have the following results.

Corollary 9.2.1.
If $m = 0$, $\delta_2 / \alpha > \delta_1 > \delta_2$, and (9.2.16) holds, then, we have: (1) If $x_j < 0$, $j = 1, 2$, the system has a unique equilibrium; 2) If $x_j > 0$, $j = 1, 2$, the system has a unique equilibrium; 3) If $x_1 > 0$ and $x_2 < 0$ ($x_1 < 0$ and $x_2 > 0$), the system has either two equilibria or no equilibrium. Moreover, the equilibrium values of the variables are given by the following procedure: Λ by (9.2.15) $\to$ Γ by (9.2.8) $\to$ N_j, $j = 1, 2$, by (9.2.9) $\to$ Z by (9.2.17) $\to$ r,

w_j and K_j by (9.2.6) $\rightarrow$ $K = K_1 + K_2$ $\rightarrow$ k_j by (9.2.5) $\rightarrow$ c_j and s_j by (9.2.2) $\rightarrow$ F_j by (9.1.1) $\rightarrow$ y_j by (9.1.3) $\rightarrow$ U_j by (9.1.4).

The requirement, $m = 0$, means that the two regions have the same level of knowledge utilization efficiency. The requirement, $\delta_1 > \delta_2$ (i.e., $\lambda_2 > \lambda_1$), implies that region 2's propensity to hold wealth is higher than region 1. The condition, $\delta_2 / \alpha > \delta_1 > \delta_2$, implies that the preference difference between the two regions is not large. The condition, (9.2.16), is guaranteed if the amenity ratio between the two regions is properly bound. As m_j is region j's knowledge utilization efficiency parameter and ε_j denotes the return to scale effects of knowledge in knowledge accumulation, we may interpret x_j as the measurement of return to scale effects of knowledge in region j. The condition, $x_j < (>)\ 0$, $j = 1, 2$, may be interpreted as that the two regions exhibit weak (strong) return to scale effects of knowledge.

In the remainder of this chapter, we examine effects of changes in some parameters on the economic structure. For simplicity, we assume that the assumptions in Corollary 9.2.1 are satisfied. Moreover, we are concerned with the two cases, $x_j > 0$ and $x_j < 0$, $j = 1, 2$, that guarantee the existence of a unique equilibrium in the dynamic system.

9.3 The Impact of Changes in Some Parameters

This section is concerned effects of changes in some parameters on the economic equilibrium structure. First, we are concerned with the impact of changes in the parameter, A. By the definition of A, A is changed either due to changes in A_j or λ_j. Here, we assume that a change in A is due to changes in amenity levels. An increase in A means an increase in A_2 or decrease in A_1.

Taking derivatives of (9.2.8), (9.2.9), $r = \alpha / \Lambda^{\beta}$ in (9.2.6), and (9.2.15) with respect to A yields

$$\frac{\beta}{\alpha \Lambda^{\alpha}} \frac{d\Lambda}{dA} = \frac{\delta_1 - \delta_2}{(\delta_1 A - \delta_2)^2} > 0, \quad \frac{dr}{dA} = - \frac{\alpha \beta}{\Lambda^{\beta+1}} \frac{d\Lambda}{dA} < 0,$$

$$\frac{d\Gamma}{dA} = \frac{\delta_1 - \delta_2}{(1 - \delta_2 \Lambda^{\beta})(1 - \delta_1 \Lambda^{\beta})} \frac{\beta \Gamma}{\Lambda^{\alpha}} \frac{d\Lambda}{dA} - \frac{\Gamma}{A} < 0,$$

$$\frac{1}{N_1} \frac{dN_1}{dA} = \frac{1}{(1 + \Gamma)\Gamma} \frac{d\Gamma}{dA} < 0, \quad \frac{1}{N_2} \frac{dN_2}{dA} = \frac{-1}{1 + \Gamma} \frac{d\Gamma}{dA} > 0. \tag{9.3.1}$$

As region 2 's amenity level is increased, the rate of interest is reduced and some of region 1 's population migrate to region 2. By (9.2.17), we have

$$- N \frac{d\Phi_0}{dZ} \frac{dZ}{dA} = \frac{\alpha d}{\Lambda^{\beta}} \frac{d\Lambda}{dA} + \left(\frac{\Phi_{01}}{N_1} - \frac{\Phi_{02}}{N_2} \right) \frac{N_2}{1 + \Gamma} \frac{d\Gamma}{dA}. \tag{9.3.2}$$

Since $d\Phi / dZ < (>) \, 0$ in the case of $x_j < (>) \, 0$, we conclude that if we further require $\Phi_{02} / N_2 \geq \Phi_{01} / N_1$, then $dZ / dA > (<) \, 0$ in the case of $x_j < (>) \, 0$. In the case that the contribution, Φ_{02} / N_2, to knowledge accumulation by region 2 's per capita is larger than the contribution, Φ_{01} / N_1, by region 1 's per capita, if the two regions exhibit strong (weak) increasing return to scale effects in knowledge utilization and accumulation, then an improvement in region 2 's amenity will reduce (increase) the equilibrium value, Z, of the knowledge stock. In the case of $\Phi_{02} / N_2 < \Phi_{01} / N_1$, it is difficult to determine the sign of dZ / dA.

By (9.2.6) and $F_j = K_j / \Lambda^{\beta}$, we get the impact on K_j and w_j as follows

$$\frac{1}{w_j} \frac{dw_j}{dA} = \frac{m_1}{\beta Z} \frac{dZ}{dA} + \frac{\alpha}{\Lambda} \frac{d\Lambda}{dA}, \quad j = 1, 2,$$

$$\frac{1}{K_j} \frac{dK_j}{dA} = \frac{1}{N_j} \frac{dN_j}{dA} + \frac{1}{\Lambda} \frac{d\Lambda}{dA} + \frac{m_j}{\beta Z} \frac{dZ}{dA},$$

$$\frac{1}{F_j} \frac{dF_j}{dA} = \frac{1}{K_j} \frac{dK_j}{dA} - \frac{\beta}{\Lambda} \frac{d\Lambda}{dA}. \tag{9.3.3}$$

In the case of $dZ / dA > 0$, the sign of dw_j / dA is the same as that of dZ / dA. We see that the two regions' wage rates may be either increased or reduced in the case of $dZ / dA < 0$. From (9.3.2) and (9.3.3), we see that it is complicated to explicitly determine the signs of dK_j / dA and dF_j / dA.

By (9.2.5), (9.2.2) and $y_j = \delta_j k_j$, we have

$$\frac{1}{k_j}\frac{dk_j}{dA} = \frac{m_1}{\beta Z}\frac{dZ}{dA} + \frac{\alpha\delta_j - r}{(\delta_j - r)\Lambda}\frac{d\Lambda}{dA},$$

$$\frac{1}{y_j}\frac{dy_j}{dA} = \frac{1}{c_j}\frac{dc_j}{dA} = \frac{1}{k_j}\frac{dk_j}{dA}. \tag{9.3.4}$$

As $d\Lambda / dA > 0$, $\delta_j - r > 0$, $\alpha\delta_1 - r > 0$ and $\alpha\delta_2 - r > 0$, we conclude that $dk_1 / dA > 0$ ($dk_2 / dA < 0$) in the case of $dZ / dA > (<)\ 0$. The signs of dk_2 / dA (dk_1 / dA) may be either negative or positive in the case of $dZ / dA > (<)\ 0$.

We thus provided the impact of changes in the two regions' amenity level on the economic structure. The effects are dependent on the two regions' preference structures and knowledge utilization efficiency and creativity in a complicated way. We can similarly analyze effects of changes in other parameters on the economic structure. The following two propositions summarize the effects of changes in the population, N, and the two regions' knowledge accumulation efficiency parameters, τ_j, on the economic structure.

Proposition 9.3.1. (The impact of changes in N)
If $m = 0$, $\delta_2 / \alpha > \delta_1 > \delta_2$, and (9.2.16) holds, an increase in the population, N, has the following impact on the equilibrium structure: (1) The two regions' population, N_j, are increased and the rate of interest, r, is not affected; (2) The level, Z, of knowledge stocks and the wage rates, w_j, in the two regions are increased (reduced) in the case of $x_j < (>)\ 0$; (3) The levels, K_j, of the capital stocks employed by the two regions and the levels, F_j, of the two regions' output are increased (may be either increased or reduced) in the case of $x_j < (>)\ 0$; (4) The net incomes, y_j, of per capita in the two regions and the levels, k_j and c_j, of capital stocks owned by per capita and per capita's consumption, in the two regions are increased (reduced) in the case of $x_j < (>)\ 0$.

Proposition 9.3.2 (The impact of changes in τ_j)

If $m = 0, \delta_2 / \alpha > \delta_1 > \delta_2$, and (9.2.16) holds, an increase in region j 's knowledge accumulation efficiency, τ_j, has the following impact on the equilibrium structure: (1) The two regions' population, N_j, and the rate of interest, r, are not affected; (2) The level, Z, of knowledge stocks and the wage rates, w_j, in the two regions are increased (reduced) in the case of $x_j < (>) \, 0$; (3) The levels, K_j, of the capital stocks employed by the two regions and the levels, F_j, of the two regions' output are increased (reduced) in the case of $x_j < (>) \, 0$; (4) The net incomes, y_j, of per capita in the two regions and the levels, k_j and c_j, of capital stocks owned by per capita and per capita's consumption, in the two regions are increased (reduced) in the case of $x_j < (>) \, 0$.

We prove the above two propositions in the appendix.

9.4 On Knowledge and Regional Development

We proposed a compact two-region economic model with endogenous capital and knowledge accumulation. Each region is characterized by its creativity, knowledge utilization, amenity and preference. The model describes an interaction between capital accumulation, knowledge growth, regional distribution of capital and labor, and locational prices over time under perfectly competitive institution. We provided the conditions for the existence of equilibria. We analyzed effects of changes in some parameters on the economic equilibrium structure.

Irrespective of its some strict assumptions, the model is structurally general in the sense that it synthesizes the important ideas in some well-established models, such as the Solow-Swan one-sector neoclassical growth, the Arrow's learning by doing model, and the neoclassical trade model with endogenous capital. For illustration, we analyzed the behavior of system under some strict assumptions. The regional economic structure is dependent on the combination of different forces such as propensities to hold wealth, knowledge accumulation and utilization efficiency and regional amenities. We may further examine dynamic behavior of the model, for instance, by simulation. Although we defined a dynamic growth model, our analysis was mainly concerned with the equilibrium structure of the economic system. It is significant to simulate the model to see how regional amenities and other factors may affect dynamic features, such as regional growth rate at any point of time and regional allocation of workers, of the economic system. We may extend the model in different ways. For instance, it is significant to introduce multiple sectors into the regional economy. Another significant direction is to make amenity levels as an endogenous variable. Pollution levels and regional infrastructure are strongly related to regional economic development.

Appendix

A.9.1 Proving Proposition 9.3.1

Taking the derivatives of the equations with respect to N, we get

$$\frac{d\Gamma}{dN} = \frac{d\Lambda}{dN} = \frac{dr}{dN} = 0, \quad \frac{dN_j}{dN} = \frac{N_j}{N}, \tag{9.A.1.1}$$

$$-N\frac{d\Phi_0}{dZ}\frac{dZ}{dN} = \Phi_{01} + \Phi_{02} > 0, \tag{9.A.1.2}$$

$$\frac{1}{w_j}\frac{dw_j}{dN} = \frac{m_1}{\beta Z}\frac{dZ}{dN}, \quad j = 1, 2,$$

$$\frac{1}{K_j}\frac{dK_j}{dN} = \frac{1}{F_j}\frac{dF_j}{dN} = \frac{1}{N} + \frac{m_1}{\beta Z}\frac{dZ}{dN}, \tag{9.A.1.3}$$

$$\frac{1}{y_j}\frac{dy_j}{dN} = \frac{1}{c_j}\frac{dc_j}{dN} = \frac{1}{k_j}\frac{dk_j}{dN} = \frac{1}{w_j}\frac{dw_j}{dN}. \tag{9.A.1.4}$$

A.9.2 Proving Proposition 9.3.2

Taking the derivatives of the equations with respect to τ_1, we get

$$\frac{d\Gamma}{d\tau_1} = \frac{d\Lambda}{d\tau_1} = \frac{dr}{d\tau_1} = \frac{dN_j}{d\tau_1} = 0, \tag{9.A.2.1}$$

$$-N\frac{d\Phi_0}{dZ}\frac{dZ}{d\tau_1} = \frac{\Phi_{01}}{\tau_1} > 0, \tag{9.A.2.2}$$

$$\frac{1}{w_j}\frac{dw_j}{d\tau_1} = \frac{m_1}{\beta Z}\frac{dZ}{d\tau_1}, \quad j = 1, 2,$$

$$\frac{1}{K_j}\frac{dK_j}{d\tau_1} = \frac{1}{F_j}\frac{dF_j}{d\tau_1} = \frac{m_1}{\beta Z}\frac{dZ}{d\tau_1}, \tag{9.A.2.3}$$

$$\frac{1}{y_j}\frac{dy_j}{d\tau_1} = \frac{1}{c_j}\frac{dc_j}{d\tau_1} = \frac{1}{k_j}\frac{dk_j}{d\tau_1} = \frac{1}{w_j}\frac{dw_j}{d\tau_1}. \tag{9.A.2.4}$$

10 Regional Economic Structure with Endogenous Knowledge

Although Adam Smith held that a country could gain from free trade, he failed to create a convincing economic theory of interregional trade. It is generally agreed that David Ricardo is the creator of the classical theory of trade. The theories of comparative advantage and the gains from trade are usually connected with Ricardo. In this theory the crucial variable used to explain international trade patterns is technology. The theory holds that a difference in comparative costs of production is the necessary condition for the existence of international trade. But this difference reflects a difference in the techniques of production. According to this theory, technological differences between countries determine the international division of labor and consumption and trade patterns. It concludes that trade is beneficial to all participating countries. This conclusion is in contradiction to the viewpoint about trade held by the doctrine of mercantilism. In mercantilism regulation and planning of economic activity are held to be efficient means of fostering the goals of a nation. This theory assumes that production costs are independent from factor prices and the composition of output. Since it lacks consideration of the demand side, the theory cannot determine how and at what value the terms of trade are or should be. The Ricardian theory failed to determine the terms of trade, even though it can be used to determine the limits in which the terms of trade must lie. It was recognized long ago that in order to determine the terms of trade, it is necessary to build trade theory that not only takes account of the productive side but also the demand side.

The Ricardian model and Heckscher-Ohlin model are two basic models of trade and production. They provide the pillars upon which much of pure theory of international trade rests. The so-called Heckscher-Ohlin model has been one of the dominant models of comparative advantage in modern economics. The Heckscher-Ohlin theory emphasizes the differences between the factor endowments of different countries and differences between commodities in the intensities with which they use these factors. The basic model deals with a long-term general equilibrium in which the two factors are both mobile between sectors and the cause of trade is different between countries with different relative factor endowments. This theory deals with the impact of trade on factor use and factor rewards. The theory is different from the Ricardian model that isolates differences in technology between countries as the basis for trade. In the Heckscher-Ohlin theory costs of production are endogenous in the sense that they are different in the trade and autarky situations, even when all countries have access to the same technology for producing each good. This model has been a main stream of international trade theory. The Heckscher-Ohlin theory provides simple, intuitive insights into the relationships between commodity prices and factor prices, factor supplies and factor rewards, and factor endowments and the pattern of production and trade. It is well known that one topic that was almost entirely absent from the pure theory of international trade was any consideration of the connection between economic growth and international trade in classical literature of economic theory. Almost all trade models developed before the 1960s are static in the sense that the supplies of factors of production are given and do not vary over time; the classical Ricardian theory of comparative advantage and the Heckscher-Ohlin theory are static since labor and capital stocks (or land) are assumed to be given and constant over time. Although Marshall held that it is important to study international trade in order to be clear of the causes which determine the economic progress of nations, it has only been in the last three or four decades that trade theory has made some systematic treatment of endogenous capital accumulation or technological change in the context of international economics.

Irrespective of many efforts on international trade (Zhang, 1992a, 1994b, 1994c), there were only a few formal economic models which deal with interdependence between interregional economic growth and knowledge accumulation. Many interregional aspects remain to be further examined. This chapter combines the main ideas of the previous two chapters. We propose a two-region and multi-sector model of perfect competition with endogenous knowledge to examine the role of knowledge in the interregional division of labor and consumption. In addition to explicitly modeling endogenous knowledge, we also take account of some other important aspects of economic geography. We emphasize the geographical character of services. This chapter is organized as follows. Section 10.1 defines a two-region and multi-sector economic model with endogenous knowledge accumulation. Section 10.2 shows that a dynamic system may have either a unique or multiple equilibria and each equilibrium may be either stable or unstable, depending upon knowledge utilization and creativity of the different sectors. Section 10.3 examines the effects of changes in the efficiency of knowledge accumulation and amenities upon the equilibrium economic geography. Section 10.4 concludes the study. The appendix offers the proof of the results presented in Section 10.2.

10.1 The Dynamics of Regional Economic Structure

To treat economic geography as an endogenous and contextually determined entity, we concentrate our study on an isolated state. This method was initiated by von Thünen (1826). The isolated state consists of two regions, indexed by 1 and 2, respectively. The system produces two commodities, indexed by 1 and 2, respectively. Each region is assumed to produce only one commodity. Each region has two production sectors: industry and services. Services are produced by combining knowledge, labor and land. Industrial production is regionally specified. Region j produces commodity j. It is assumed that each region's product is homogeneous. We assume a homogenous and fixed national labor force, N. We neglect any cost for migration and professional changes. We assume perfect competition in all markets.

Each region has fixed land. The land is distributed between the service sector and housing by perfect competition. We select commodity 1 to serve as numeraire, with all the other prices being measured relative to its price. For simplicity of analysis, we neglect transportation cost of commodities between and within regions. It can be shown that we may introduce transportation costs for commodities by taking Samuelson's iceberg form, in which transport costs are incurred in the good transported. It should be noted that by adding transportation costs some of our results on geographical pattern formation may become stronger. As our model is already too complicated, we will assume zero transportation costs at this initial stage. The assumption of zero transportation cost of commodities implies price equality for any commodities between the two regions. As amenity and land are immobile, wage rates and land rent are not necessarily identical between the two regions.

We introduce

i, s —subscript index for industry and services, respectively;

N and L_j —the fixed total population and region j 's territory size, $j = 1, 2$;

$F_{ij}(t)$ and $F_{sj}(t)$ — the output levels of region j 's industrial and service sectors at time t, $j = 1, 2$;

$N_{ij}(t)$ and $N_{sj}(t)$ —the labor force employed by region j 's industrial and the service sectors;

$N_j(t)$ —region j 's population j;

$L_{sj}(t)$ and $L_{hj}(t)$ — the land size used by the service sector and for housing in region j;

$Z(t)$ —the knowledge stock of the national economy;

$p(t)$ and $p_j(t)$ —price of commodity 2 and price of services in region j; and

$w_j(t)$ and $R_j(t)$ —region j 's wage rate and land rent in region j.

Production sectors

First, we describe behavior of the production sectors in the system. We assume that service production is to combine knowledge, labor and land. We specify the production functions of the service sectors as follows

$$F_{sj} = Z^{m_{sj}} L_{sj}^{\alpha} N_{sj}^{\beta}, \quad \alpha, \beta > 0, \quad \alpha + \beta = 1, \quad m_{sj} \geq 0, \quad j = 1, 2. \tag{10.1.1}$$

We only use labor and knowledge as input factors. It should be noted that to introduce other factors such as capital may cause great analytical difficulties. We use variable $Z(t)$ to measure, in an aggregated sense, stock of knowledge in the society at time t. In this chapter, knowledge is treated as public good in the sense that utilization of knowledge by any agent in the system will not affect that by any other. We assume that knowledge utilization efficiency varies spatially and professionally. The parameter, m_{sj}, describes knowledge utilization efficiency. We call m_{sj} the knowledge utilization efficiency parameter of region j's service sector. There are obviously limitations to knowledge as public goods in modeling the innovation and imitation processes. But to model the complexity of private and public characteristics of knowledge, it is necessary to further disaggregate the knowledge variable into multiple components. This will result in high-dimensional dynamic problems.

The marginal conditions of service production are given by

$$R_j = \frac{\alpha p_j F_{sj}}{L_{sj}}, \quad w_j = \frac{\beta p_j F_{sj}}{N_{sj}}, \quad j = 1, 2. \tag{10.1.2}$$

We specify the two regions' industrial production functions as follows

$$F_{ij} = Z^{m_{ij}} N_{ij}, \quad m_{ij} \geq 0, \quad j = 1, 2 \tag{10.1.3}$$

where m_{ij} are the knowledge utilization efficiency parameter of region j's industrial sector.

The conditions of zero-profit are given by

$$w_1 = Z^{m_{i1}}, \quad w_2 = pZ^{m_{i2}}. \tag{10.1.4}$$

Consumers' behavior

We assume that a household's utility in a given region is dependent on the amenity level, per household's consumption levels of industrial commodities and housing conditions. We measure housing conditions by lot size. The utility function is specified as follows

$$U_j(t) = \frac{A_j C_{sj}^{\gamma} C_{1j}^{\xi_1} C_{2j}^{\xi_2} L_{hj}^{\eta}}{N_j}, \quad j = 1, 2, \quad \gamma, \xi_1, \xi_2, \eta > 0,$$

$$\gamma + \xi_1 + \xi_2 + \eta = 1 \tag{10.1.5}$$

in which $C_{sj}(t)$, $C_{1j}(t)$, and $C_{sj}(t)$ are respectively region j's consumption levels of services, commodity 1 and commodity 2 at time t. In (10.1.5), A_j denotes region j's amenity level. Amenity is an aggregated measure of regional livening conditions such as infrastructure status, regional culture and climate. Some locational amenities such as pollution level, residential density and transportation congestion may be dependent on economic agents' activities, while other locational amenities such as climate, transport structure, and historical buildings, may not be strongly affected by economic agents' activities, at least, within the short-run. Accordingly, in a strict sense, it is necessary to classify amenities into endogenous and exogenous categories when we deal with dynamics of economic geography. Which kinds of amenities should be classified as endogenous or exogenous also depends upon the time scale of the analysis and the economic system under consideration.

The assumption that utility level is identical over space at any point of time is represented by

$$U_1(t) = U_2(t). \tag{10.1.6}$$

The consumer problem is defined by

$$\max U_j, \ s.t.: p_j C_{sj} + C_{1j} + pC_{2j} + R_j L_{hj} = w_j N_j. \tag{10.1.7}$$

We have the following optimal solutions

$$C_{sj} = \frac{\gamma w_j N_j}{p_j}, \ C_{1j} = \xi_1 w_j N_j, \ C_{2j} = \frac{\xi_2 w_j N_j}{p}, \ L_{hj} = \frac{\eta w_j N_j}{R_j}, \quad j = 1, 2. \tag{10.1.8}$$

The balances of demand for and supply of commodities are given by

$$C_{j1} + C_{j2} = F_{ij}, \ C_{sj} = F_{sj}, \ j = 1, 2. \tag{10.1.9}$$

Full employment of labor and land

The assumption that labor force and land are fully employed is represented by

$$N_{ij}(t) + N_{sj}(t) = N_j(t), \ N_1(t) + N_2(t) = N, \ L_{sj}(t) + L_{hj}(t) = L_j, \quad j = 1, 2. \tag{10.1.10}$$

Knowledge accumulation

Similar to Zhang (1992b, 1993d), we model knowledge accumulation as follows

$$\frac{dZ}{dt} = \sum_{j=1}^{2}\left(\frac{\tau_{sj}F_{sj}}{Z^{\varepsilon_{sj}}} + \frac{\tau_{ij}F_{ij}}{Z^{\varepsilon_{ij}}}\right) - \delta_z Z \tag{10.1.11}$$

in which δ_z is the fixed depreciation rate of knowledge, and τ_{sj} (≥ 0), τ_{ij} (≥ 0), ε_{sj} and ε_{ij}, $j = 1, 2$, are parameters.

We only take account of learning by doing effects in knowledge accumulation. The term, $\tau_{i1}F_{i1} / Z^{\varepsilon_{i1}}$, for instance, implies that contribution of region 1's industrial sector to knowledge is positively related to its production scale, F_{i1}, and is dependent on the current level of knowledge stocks. The term $Z^{\varepsilon_{i1}}$ takes account of returns to scale effects in knowledge accumulation. The case of $\varepsilon_{i1} > (<) 0$ implies increasing (decreasing) returns to scale in knowledge accumulation. The other three terms in (10.1.11) can be similarly interpreted. Knowledge accumulation may be affected by many factors in different ways. The above specification takes account of one source of knowledge accumulation. We omit possible effects of R&D activities on knowledge accumulation.

We built the model. The system has 30 variables, N_{sj}, N_{ij}, L_{hj}, L_{sj}, F_{sj}, F_{ij}, C_{sj}, C_{1j}, C_{2j}, U_j, N_j, w_j, p_j, R_j ($j = 1, 2$), p and Z. It contains the same number of independent equations. We now examine properties of the dynamic system.

10.2 The Equilibria Structure and Stability

This section is concerned with the conditions for the existence of economic equilibria and for stability. First, we show that for any given knowledge stock, $Z(t)$, the division of labor and economic geography are uniquely determined at any specific time. The following proposition is proved in the appendix.

Proposition 10.2.1.
For any given level of knowledge, $Z(t)$, all the other variables in the system are uniquely determined at any point of time. The values of the variables are given as functions of $Z(t)$ by the following procedure: p by (10.A.1.16) $\rightarrow$ N_j, $j = 1, 2$, by (10.A.1.5) $\rightarrow$ N_{sj} and N_{ij} by (10.A.1.1) $\rightarrow$ w_j by (10.A.1.7) $\rightarrow$ R_j by (10.A.1.8) $\rightarrow$ L_{hj} and L_{sj} by (10.A.1.10) $\rightarrow$ p_j by (10.A.1.11) $\rightarrow$

F_{sj} by (10.1.1) $\rightarrow$ $C_{sj} = F_{sj}$ $\rightarrow$ L_{ij} by (10.1.3) $\rightarrow$ $C_{11} = \xi_1 w_1 N_1$ $\rightarrow$ C_{12} by (10.A.1.6) $\rightarrow$ U_j by (10.1.5).

The above proposition guarantees that if we can find knowledge $Z(t)$, then we can explicitly solve all the other variables as functions of $Z(t)$ in the system at any point of time. Hence, to examine the dynamic properties of the system, it is sufficient to examine the dynamic properties of (10.1.11). By the procedure given in Proposition 10.2.1 we can represent $N_{sj}(t)$ and $N_{ij}(t)$ as functions of $Z(t)$ as follows

$$N_{s1} = \gamma\beta\xi_1 ANZ^m\Psi, \quad N_{i1} = (1 - \gamma\beta)\xi_1 ANZ^m\Psi,$$
$$N_{s2} = \gamma\xi_2\beta N\Psi, \quad N_{i2} = (1 - \gamma\beta)\xi_2 N\Psi \tag{10.2.1}$$

in which

$$A \equiv \left\{\left(\frac{\xi_2 L_1}{\xi_1 L_2}\right)^{\eta+\alpha\gamma}\frac{A_1}{A_2}\right\}^{\frac{1}{1-\gamma\beta}}, \quad \Psi(Z) \equiv \frac{1}{\xi_2 + \xi_1 AZ^m}, \quad m \equiv -\frac{m_s\gamma}{1-\gamma\beta},$$
$$m_s \equiv m_{s2} - m_{s1}.$$

Substituting these into the production functions (10.1.1) and (10.1.3), we get F_{ij} and F_{sj} as functions of Z. Substituting $F_{ij}(t)$ and $F_{sj}(t)$ into (10.1.11), we determine knowledge accumulation dynamics $Z(t)$, as follows

$$\frac{dZ}{dt} = \sum_{j=1}^{2}\{\Omega_{sj}(Z) + \Omega_{ij}(Z)\}Z - \delta_z Z \tag{10.2.2}$$

where

$$\Omega_{s1} = \tau_{s1}(\beta\xi_1\xi_2 AN)^\beta L_{s1}^\alpha Z^{x_{s1}}\Psi^\beta, \quad \Omega_{i1} = v_1 Z^{x_{i1}}\Psi,$$
$$\Omega_{s2} = \tau_{s2}(\gamma\beta\xi_2 N)^\beta L_{s2}^\alpha Z^{x_{s2}}\Psi^\beta, \quad \Omega_{i2} = v_2 Z^{x_{s2}}\Psi \tag{10.2.3}$$

in which

$$v_1 \equiv (1 - \gamma\beta)\tau_{i1}\xi_1 AN>0, \quad v_2 \equiv (1 - \gamma\beta)\tau_{i2}\xi_2 N>0,$$
$$x_{s1} \equiv m_{s1} - m - \varepsilon_{s1} - 1, \quad x_{s2} \equiv m_{s2} - \varepsilon_{s2} - 1,$$

$$x_{i1} \equiv m_{i1} - m - \varepsilon_{i1} - 1, \quad x_{i2} \equiv m_{i2} - \varepsilon_{i2} - 1. \tag{10.2.4}$$

For simplicity of discussion, we make the following assumption.

Assumption 10.2.1.
In the remainder of this chapter, we require $\tau_{s1} = \tau_{s2} = 0$.

This requirement implies $\Omega_{s1} = \Omega_{s2} = 0$. The service sectors make no contribution to knowledge accumulation. From the functional forms of Ω_{sj} and Ω_{ij} in (10.2.2) and (10.2.3) we see that the omission of service sectors' creativity does not affect essential conclusions of our model. If we don't make this omission, we have to discuss more cases of different combinations of the parameters. As these discussions will provide little new insight, we limit our analysis to the special case of $\Omega_{s1} = \Omega_{s2} = 0$.

First, we notice that the dynamic properties of the system are determined by the combination of the parameters, x_{i1}, x_{i2}, and m. The three parameters may be either positive or negative. As

$$m = -\frac{m_{s2} - m_{s1}}{1 - \gamma\beta}\gamma.$$

we see that if region 1's service sector utilizes knowledge more (less) effectively than region 2's service sector, i.e., $m_{s1} > m_{s2}$ ($m_{s1} < m_{s2}$), then $m > 0$ ($m < 0$); if the two regions' knowledge utilization efficiency is equal, then $m = 0$. From $x_{i2} \equiv m_{i2} - \varepsilon_{i2} - 1$, if region 2's industrial sector is effective in knowledge utilization (i.e., m_{i2} being large) and its contribution to knowledge exhibits increasing returns to scale (i.e., $m_{i2} < 0$), then x_{i2} may be positive; if region 2's industrial sector is not effective in knowledge utilization and its contribution to knowledge exhibits decreasing returns to scale, then $x_{i2} < 0$. We can similarly discuss the sign of $x_{i1} \equiv m_{i1} - m - \varepsilon_{i2} - 1$. We say that when $x_{ij} > (<) \, 0$, region j's industrial sector exhibits increasing (decreasing) returns to scale effects in knowledge accumulation.

We can thus conclude that the parameters x_{ij} and m may be either positive or negative, depending upon various combinations of knowledge utilization efficiency and creativity of different economic sectors in the two regions. The following proposition shows that the properties of our one-dimensional differential equation, (10.2.2), are dependent on how these knowledge parameters are combined.

Proposition 10.2.2.
(1) If $x_{i1} > 0$, $x_{i2} > 0$ and $m < 0$, the system has a unique unstable equilibrium; (2) $x_{i1} < 0$, $x_{i2} < 0$ and $m > 0$, the system has a unique stable equilibrium; and (3) In each of the other six combinations of $x_{i1} < 0$ or $x_{i1} > 0$, $x_{i2} < 0$ or $x_{i2} > 0$ and $m < 0$ or $m > 0$, the system has either no equilibrium or multiple ones.

The above proposition is proved in the appendix. We interpreted the meanings of x_{i1}, x_{i2} and m. From these discussions we can directly interpret the conditions in the above proposition as implying that if the two regions exhibit increasing (decreasing) scale effects in knowledge accumulation, the system has an unstable (stable) unique equilibrium; if one region exhibits increasing scale effects in knowledge accumulation but the other region decreasing ones, the system has two equilibria.

From Propositions 3.1 and 3.2, we see that for given parameter values, we can explicitly determine the properties of the dynamic system. For simplicity, we were only concerned with the cases that none of the parameters, x_{i1}, x_{i2} and m, is equal to zero. It is not difficult to discuss the cases that one or two of the three parameters are equal to zero. The following corollary, summarizing the properties of the dynamic system in the case of $m = 0$, is proved in the appendix.

Corollary 10.2.1.
If $m = 0$, we have (1) if $x_{ij} > 0$, $j = 1, 2$, the system has a unique unstable equilibrium; (2) if $x_{ij} < 0$, $j = 1, 2$, the system has a unique stable equilibrium; and (3) if $x_{i1} < 0$ and $x_{i2} > 0$ (or $x_{i1} > 0$ and $x_{i2} < 0$), the system has two equilibria - the one with higher value of knowledge is unstable and the other one is stable.

In the remainder of this chapter, for illustration, we will examine the impact of changes in some parameters on the long-term equilibrium structure of the economic geography.

10.3 The Creativity and the Regional Economic Structure

This section is concerned with the impact of changes in knowledge accumulation efficiency parameters, τ_{ij}, $j = 1, 2$, of the industrial sector in region j on the system.

First, from (10.A.2.2) we get the impact of changes in τ_{ij} on Z as follows

$$-\Phi^* \frac{dZ}{d\tau_{ij}} = \frac{\Phi_j}{\tau_{ij}} > 0, \quad j = 1, 2 \tag{10.3.1}$$

where $\Phi^{*\prime}$ is defined in (10.A.2.3). The sign of $\Phi^{*\prime}$ may be either positive or negative, depending on the parameter values of x_{ij}. In the case of $m = 0$, we directly have the following three cases: (i) if $x_{ij} > 0$, $j = 1, 2$, $\Phi^{*\prime} > 0$ at the unique equilibrium; (ii) if $x_{ij} < 0$, $j = 1, 2$, $\Phi^{*\prime} < 0$ at the unique equilibrium; (iii) if $x_{i1} < 0$ and $x_{i2} > 0$ (or $x_{i1} > 0$ and $x_{i2} < 0$), $\Phi^{*\prime} > (<)\ 0$ at the equilibrium with the high (low) value of Z. We thus conclude that the sign of Φ' is dependent on whether each industrial sector exhibits increasing or decreasing returns. In case (i), we have $dZ / d\tau_{ij} < 0$. In case (ii), we have $dZ / d\tau_{ij} > 0$. In case (iii), we have $dZ / d\tau_{ij} <(>)\ 0$ at the equilibrium with high (low) values of Z. We see that an improvement in knowledge accumulation efficiency may either increase or decrease the equilibrium value of knowledge. For instance, in case (ii) which means that the two industrial sectors exhibit decreasing returns, an improvement in knowledge accumulation efficiency of any region's industrial sector increases the equilibrium level of knowledge. In case (i), the impact in changes of knowledge accumulation efficiency is the opposite to that in the case of stability. We see that the effects of changes in the parameters on the system are significantly dependent upon the stability conditions. To explain how the new equilibrium given by (10.3.1) is achieved through a dynamic process, we have to examine all the relations connecting the 30 variables in the system. We omit the explanation of (10.3.1) in detail.

From (10.A.1.16) we directly have

$$\frac{Z}{p}\frac{dp}{d\tau_{ij}} = (m - m_i)\frac{dZ}{d\tau_{ij}} \tag{10.3.2}$$

in which $m \geq 0$. In the remainder of this section, we assume: $x_{ij} < 0$, $j = 1, 2$. We have: $dZ / d\tau_{ij} > 0$. If $m_i < 0$, i.e., $m_{i1} > m_{i2}$, the price of commodity 2 is increased. This implies that when region 1's two sectors utilize knowledge more effectively than the two sectors in region 2, then an increase in knowledge increases the price level of commodity 2. In the case of $m_i > 0$, from (10.3.2) $dp / d\tau_{ij}$ may be either positive or negative.

Taking derivatives of (10.A.1.5) and (10.A.1.1) with respect to τ_{ij} yields

$$\frac{Z}{N_1}\frac{dN_1}{d\tau_{ij}} = m\xi_2 \frac{dZ}{d\tau_{ij}} > 0, \quad \frac{dN_2}{d\tau_{ij}} = -\frac{dN_1}{d\tau_{ij}} < 0,$$
$$\frac{1}{N_{sk}}\frac{dN_{sk}}{d\tau_{ij}} = \frac{1}{N_{ik}}\frac{dN_{ik}}{d\tau_{ij}} = \frac{1}{N_k}\frac{dN_k}{d\tau_{ij}}, \quad k = 1, 2. \tag{10.3.3}$$

The direction of migration due to the improvement in productivity is only determined by the sign of $m_{s1} - m_{s2}$. As we are only concerned with the case of $m_{s1} > m_{s2}$, we see that some of region 2's population migrate to region 1. If the knowledge utilization efficiency in the two regions' service sectors is identical, then improved productivity has no impact on the labor distribution. The labor employed by the two sectors in each region will be increased or decreased if the region's total employment is increased or decreased.

By (10.A.1.7), we get the impact on the wage rates of the two regions as

$$\frac{Z}{w_1}\frac{dw_1}{d\tau_{ij}} = m_{i1}\frac{dZ}{d\tau_{ij}} > 0, \quad \frac{Z}{w_2}\frac{dw_2}{d\tau_{ij}} = (m + m_{i1})\frac{dZ}{d\tau_{ij}} > 0. \tag{10.3.4}$$

An improvement in knowledge accumulation efficiency increases the wage rates in the two regions. From (10.A.1.8) we get the impact on the land rent as follows

$$\frac{Z}{R_1}\frac{dR_1}{d\tau_{ij}} = (m_{i1} + m\xi_2)\frac{dZ}{d\tau_{ij}} > 0,$$
$$\frac{Z}{R_2}\frac{dR_2}{d\tau_{ij}} = (m + m_{i1} - m\xi_2\frac{N_1}{N_2})\frac{dZ}{d\tau_{ij}} > 0. \tag{10.3.5}$$

We see that region 1's land rent is increased. If $N_1 / N_2 \leq 1$, then region 2's land rent is increased. But if the population of region 2 is larger than region 1, then it is possible for region 2's land rent to decline as more people move from region 2 to region 1. We see that in the case that region 2 employs more workers, the sign of $dR_2 / d\tau_{ij}$ may be either positive or negative, depending on the structure of the whole system. From (10.A.1.11), we get the effects of changes in τ_{ij} on the prices of services as follows

$$\frac{Z}{p_1}\frac{dp_1}{d\tau_{ij}} = (m_{i1} + \alpha m\xi_2 - m_{s1})\frac{dZ}{d\tau_{ij}},$$

$$\frac{Z}{p_2}\frac{dp_2}{d\tau_{ij}} = (m + m_{i1} - m_{s2} - \alpha m\xi_2 \frac{N_1}{N_2})\frac{dZ}{d\tau_{ij}} > 0. \qquad (10.3.6)$$

We see that as knowledge accumulation efficiency is improved, prices of services may be either increasing or decreasing. From the above discussion, it is easy to get the impact of changes in τ_{ij} upon all the 30 variables in the system. It can be seen that the effects are very sensitive to how different values of the parameters are combined.

As we have explicitly solved the equilibrium problem, it is straightforward to examine the effects of changes in any parameter in the system as we did for τ_{ij}. For instance, by taking derivatives of (10.A.2.2) with respect to A_2, we get

$$- \Phi^* \frac{dZ}{dA_2} = \frac{\xi_1 AZ^m - \Phi_1}{(1 - \gamma\beta)A_2} \qquad (10.3.7)$$

where $\Phi^{*\prime} < 1$ and $\Phi_1 > 0$. An improvement in region 2's amenity may either increase or decrease the equilibrium level of knowledge. If the term, Φ_1, associated with creativity of the industrial sector in region 1 is larger than the term, $\xi_1 AZ^m$, then dZ / dA_2 is negative. This can be interpreted as that when region 1's industrial sector has a high contribution to knowledge accumulation, then an increase in region 2's living conditions may reduce the long-run equilibrium level of knowledge. An improvement in region 2's amenity causes a change in region 1's equilibrium utility level. This will result in re-allocation of labor force. As knowledge utilization and creativity are spatially different, the re-allocation of labor force will cause shifts in the equilibrium structure. It can be seen from the above equation that whether an improvement in amenity has a positive or negative impact on the equilibrium level of knowledge stocks is dependent on a complicated interdependence between many factors.

There are two important points. One is that if we take derivatives of the equilibrium condition for knowledge accumulation with respect to region 1's amenity, A_1, then we can show that the sign of dZ / dA_1 is identical to that of $\Phi_1 - \xi_1 AZ^m$. The other one is that if we are concerned with the case that the system has an unstable unique equilibrium, i.e., $x_{ij} > 0$, $j = 1, 2$, and $m \le 0$, then $dZ / dA_2 < 0$, which is opposite to the case that the system has a unique stable equilibrium.

By (10.A.1.16), we have

$$\frac{Z}{p}\frac{dp}{dA_2} = - \frac{Z}{(1 - \gamma\beta)A_2} + (m - m_i)\frac{dZ}{dA_2}. \tag{10.3.8}$$

To see how the price of commodity 2 is changed by changes in A_2, we first look at (10.1.6). As region 2's amenity is changed, the balance condition (10.1.6) is disturbed. Therefore, migration between the two regions will be necessary for the system to achieve a new equilibrium (because we assume the existence of a unique stable equilibrium). As the employment distribution is changed, the demand conditions (10.1.8) will be disturbed, which will result in adjustment of the price variables. As the prices and demand conditions are changed, the production scales will be shifted toward the new equilibrium. All of these changes further shift the balance condition (10.1.6). The process will thus be repeated until the system achieves at the new equilibrium. The shift in the price of commodity 2 is given by (10.3.7). The new price level may become either higher or lower than the old one.

From (10.A.1.5) we get the impact on N_j as follows

$$\frac{\Psi}{\xi_1 AN_2Z^m}\frac{dN_2}{dA_2} = \frac{1}{(1 - \gamma\beta)A_2} - \frac{m}{Z}\frac{dZ}{dA_2}, \quad \frac{dN_1}{dA_2} = - \frac{dN_2}{dA_2}. \tag{10.3.9}$$

If the knowledge utilization efficiency of the two service sectors is identical, i.e., $m = 0$, then an improvement in region 2's amenities attracts more people from region 1 to region 2. But in the case that $m > 0$ and dZ / dA_2 is large, an improvement in region 2's amenities may not increase region 2's population. Hence, the direction of migration is also dependent on the characteristics of knowledge utilization in the two regions. We may easily provide the effects of changes in A_2 on the other variables in the system. As little new insight can be gained, we omit further examination.

10.4 Regional Dynamics with Knowledge

This chapter proposed a perfectly competitive regional economic dynamic model with endogenous knowledge accumulation. We showed that the dynamic system may have either a unique or multiple equilibria and each equilibrium may be either stable or unstable, depending upon the knowledge utilization efficiency and creativity of the various sectors. We also examined the effects of changes in knowledge accumulation efficiency and regional amenity on the equilibrium structure. The model may be extended in different ways. For instance, we may enrich the spatial structure of the model by further specifying location of the industrial section, services and residents within each region. We may also enrich endogenous dynamic processes of the regional division of labor with transportation costs.

Appendix

Proving Proposition 10.2.1

From $C_{sj} = F_{sj}$ together with (10.1.2)

$$N_{ij} + N_{sj} = N_j, \quad C_{sj} = \frac{\gamma w_j N_j}{p_j}$$

we have

$$N_{sj} = \gamma \beta N_j, \quad N_{ij} = (1 - \gamma \beta) N_j, \quad j = 1, 2. \tag{10.A.1.1}$$

Substituting C_{ij} in (10.1.8) into $C_{j1} + C_{j2} = F_{ij}$ in (10.1.9) together with (10.1.3) and (10.1.4) yields

$$N_1 + Z^{m_i} N_2 = \frac{N_{i1}}{\xi_1}, \quad N_1 + Z^{m_i} N_2 = \frac{p Z^{m_i} N_{i2}}{\xi_2} \tag{10.A.1.2}$$

where $m_i \equiv m_{i2} - m_{i1}$. As the left-hand sides of the two equations in (10.A.1.2) are identical, we have

$$\frac{N_{i1}}{N_{i2}} = \frac{\xi_1 p Z^{m_i}}{\xi_2}. \tag{10.A.1.3}$$

The ratio of the employment in the two industrial sectors depends upon the difference in knowledge utilization efficiency and in marginal utility of the two industrial commodities. Substituting N_{ij} in (10.A.1.1) into (10.A.1.3) yields

$$\frac{N_1}{N_2} = \frac{\xi_1 p Z^{m_i}}{\xi_2}. \tag{10.A.1.4}$$

The ratio of employment between the two regions is related to the preference parameters, level of knowledge and knowledge utilization parameters. By the above equation and $N_1 + N_2 = N$, we get

$$N_1 = \xi_1 ANZ^m \Psi(Z), \quad N_2 = \xi_2 N\Psi(Z). \tag{10.A.1.5}$$

By (10.A.1.1) and (10.A.1.5), we solved the labor distribution, N_{sj}, N_{ij} and N_j, as functions of p and Z. We determine F_{ij} as functions of p and Z by substituting N_{ij} in (10.A.1.1) into (10.1.3). Region 1 's consumption of commodity 1 is given by: $C_{11} = \xi_1 w_1 N_1$. From $C_{11} + C_{12} = F_{i1}$, we get

$$C_{12} = F_{i1} - C_{11} = \xi_0 F_{i1} \tag{10.A.1.6}$$

in which $\xi_0 \equiv 1 - \xi_1 / (1 - \gamma\beta) > 0$. By (10.1.4), $C_{12} = \xi_1 w_2 N_2$ and (10.A.1.6), we have

$$w_1 = Z^{m_{i1}}, \quad w_2 = \frac{\xi_0 A(1 - \beta) Z^{m + m_{i1}}}{\xi_2}. \tag{10.A.1.7}$$

From (10.1.2) we get

$$L_{sj} = \frac{\alpha w_j N_{sj}}{\beta R_j}.$$

Substituting this and $L_{hj} = \eta w_j N_j / R_j$ and (10.A.1.1) into the last equation in (10.1.10), we obtain

$$R_j = \frac{(\eta + \alpha\gamma) w_j N_j}{L_j}, \quad j = 1, 2. \tag{10.A.1.8}$$

By

$$L_{hj}=\frac{\eta w_j N_j}{R_j}, \quad L_{sj}=\frac{\alpha w_j N_{sj}}{\beta R_j} \tag{10.A.1.9}$$

(10.A.1.8) and (10.A.1.5), we get

$$L_{hj}=\frac{\eta L_j}{\eta+\alpha\gamma}, \quad L_{sj}=\frac{\alpha\gamma L_j}{\eta+\alpha\gamma}. \tag{10.A.1.10}$$

From L_{sj} and N_{sj}, we determine F_{sj}. We obtain C_{sj} as functions of Z by $C_{sj}=F_{sj}$. Utilizing $C_{sj}=\gamma w_j N_j / p_j$ in (10.1.8), we get

$$p_j=\frac{\gamma w_j N_j}{F_{sj}}, \quad j=1,\,2. \tag{10.A.1.11}$$

We have thus shown that the 28 variables, N_{sj}, N_{ij}, L_{hj}, L_{sj}, F_{sj}, F_{ij}, C_{sj}, C_{1j}, C_{2j}, U_j, N_j, w_j, p_j, R_j ($j=1,\,2$), are determined as functions of p and Z. We now show that commodity 2's price can be determined as a function of Z.

Substituting (10.1.8) into (10.1.6) yields

$$\frac{w_1}{w_2}=\left(\frac{p_1}{p_2}\right)^{\gamma}\left(\frac{R_1}{R_2}\right)^{\eta}\frac{A_2}{A_1}. \tag{10.A.1.12}$$

The equation gives a relation between the wage rates, amenity, service prices and land rents in the two regions.

From (10.1.4) we have

$$\frac{w_1}{w_2}=\frac{1}{pZ^{m_i}}. \tag{10.A.1.13}$$

By (10.A.1.4), (10.A.1.13) and (10.A.1.8), we have

$$\frac{R_1}{R_2} = \frac{\xi_2 L_2}{\xi_1 L_1}. \tag{10.A.1.14}$$

From (10.A.1.11), (10.1.1) and (10.A.1.1), we have

$$\frac{p_1}{p_2} = \left(\frac{w_1}{w_2}\right)^{\beta}\left(\frac{R_1}{R_2}\right)^{\alpha} Z^{m_s} \tag{10.A.1.15}$$

where w_1 / w_2 is given in (10.A.1.13). Substituting (10.A.1.13)-(10.A.1.15) into (10.A.1.12) yields

$$p = AZ^{m-m_i}. \tag{10.A.1.16}$$

We have thus proved that for any given knowledge, Z, the system has a unique temporary equilibrium.

A.10. 2 Proving Proposition 10.2.2

In the case of $\Omega_{s1} = \Omega_{s2} = 0$, the knowledge accumulation is given by

$$\frac{dZ}{dt} = \{\Phi_1(Z) + \Phi_2(Z)\}Z\Psi(Z) - \delta_z Z \equiv \Phi(Z) \tag{10.A.2.1}$$

where $\Phi_j = v_j Z^{x_{ij}}$, $j = 1, 2$. We now examine properties of (10.A.2.1). Equilibrium is given as a solution of the following equation

$$\Phi^*(Z) \equiv \Phi_1(Z) + \Phi_2(Z) - \xi_1 AZ^m - \xi_2\delta_z = 0. \tag{10.A.2.2}$$

Our problem is whether or not $\Phi^*(Z) = 0$ has a positive solution.

In the case of $x_{i1} > 0$, $x_{i2} > 0$ and $m < 0$, we have: $\Phi^*(0) < 0$, $\Phi^*(+\infty) > 0$, $\Phi^{*\prime}(Z) > 0$ for $0 < Z < +\infty$. Accordingly, there is a unique positive Z such that $\Phi^*(Z) = 0$. The stability is determined by the sign of $\Phi(Z)$ at the equilibrium. If $\Phi > 0$, the system is unstable; if $\Phi' < 0$, it is stable; and if $\Phi' = 0$, it is neutral. Taking derivatives of Φ with respect to Z yields

$$\Phi' = x_{i1}\Psi\Phi_1 + x_{i2}\Psi\Phi_2 - (\Phi_1 + \Phi_2)\xi_1 m A \Psi^2 Z^{m-1} \qquad (10.A.2.3)$$

which is evaluated at equilibrium. We thus conclude the system has a unique unstable equilibrium in this case.

In the case of $x_{i1} < 0$, $x_{i2} < 0$ and $m > 0$, we have: $\Phi^*(0) > 0$, $\Phi^*(+\infty) < 0$, $\Phi^{*}{}'(Z) < 0$ for $0 < Z < +\infty$. We see that the system has a unique stable equilibrium in this case.

There are six other possible combinations of $x_{i1} < 0$ or $x_{i1} > 0$, $x_{i2} < 0$ or $x_{i2} > 0$ and $m < 0$ or $m > 0$. It can be checked that if the system has equilibrium, it must have multiple ones. For instance, in the case of $x_{i1} < 0$, $x_{i2} > 0$ and $m < 0$, because $\Phi^*(0) < 0$ and $\Phi^*(+\infty) < 0$ are held, the equation, $\Phi^*(Z) = 0$, has either no solution or multiple ones. Similarly, we can check the other five cases. We can directly check the stability of each equilibrium by (10.A.2.3). Summarizing the above discussion, we proved Proposition 10.2.2.

We now check Corollary 10.2.2. As in cases i) and ii) in Proposition 10.2.2, we can show that i) if $x_{ij} > 0$, $j = 1, 2$, then the system has a unique unstable equilibrium; ii) if $x_{ij} < 0$, $j = 1, 2$, then the system has a unique stable equilibrium. We now show that if $x_{i1} < 0$ and $x_{i2} > 0$ (or $x_{i1} > 0$ and $x_{i2} < 0$), the system has two equilibria - the one with higher value of knowledge is unstable and the other one is stable.

It is sufficient to check the case of $x_{i1} > 0$ and $x_{i2} < 0$ as the case $x_{i1} < 0$ and $x_{i2} > 0$ can be similarly proved. As $\Phi^*(0) > 0$ and $\Phi^*(+\infty) > 0$, the system has either no equilibrium or multiple ones. From $\Phi^* = (x_{i1}\Phi_1 + x_{i2}\Phi_2)/Z$, we see that $\Phi^{*}{}' = 0$ has a unique positive solution, denoted by Z_0. This implies that $\Phi^*(Z) = 0$ cannot have more than two solutions. When the system has two equilibria, the stability of each equilibrium is determined by the sign of Φ'. If $Z < Z_0$, then $\Phi' < 0$. If $Z > Z_0$, $\Phi' > 0$. We thus proved the corollary.

11 Regional Growth with Universities

This chapter is to introduce education into the interregional modeling framework proposed in the preceding chapters. It is important to examine the role of the university education and research on economic development at the regional scale. Universities play a central role in regional innovation systems as producers of basic research. The importance of basic research is due to the public good nature of the research and positive externalities to the private sector in the form of knowledge spillovers. Regional economies have experienced dramatic changes in different parts of the world in the last few decades. These changes in economic development are influenced by increases of the share of the total resources devoted to R&D, in the educational achievements of the labor force and in the rate of technological change. It has been argued that in developed economies universities and research institutions are the main actors in the knowledge generation process.

This chapter proposes a dynamic endogenous two-region growth model. One of the regions is assumed to contain a research university - the only university in the whole economy -, i.e. the model covers both the accumulation of knowledge and the accumulation of human capital. The two-region approach makes it possible to analyze how changes in human capital and knowledge accumulation affect the economic geography in various respects. It also makes it possible to analyze how changes in amenity conditions changes the location of the human capital with further effects on the general economic geography. The chapter is organised as follows. The basic model is presented in Section 11.1. In Section 11.2, we analyze the conditions for the existence of equilibria. The impacts of policies of high education are analyzed in Section 11.3; the impacts of changes in regional amenities are analyzed in Section 11.4. Section 11.5 concludes the chapter.

11.1 The Regional Model with Universities

The regional economic system of a state is assumed to consist of two regions, indexed by 1 and 2, respectively. The system produces two commodities, also indexed by 1 and 2, respectively. Each region is assumed to produce only one commodity. Each region has two production sectors, one manufacturing sector and one service sector. Production in the manufacturing sector is regionally specific. Region j produces commodity j. It is assumed that each region's product is homogeneous. Service production is achieved by combining human capital with labor. There is only one university in the system. It is located to region 1. In this chapter students are included in the total labor force available in the economy. We assume a homogeneous and fixed national labor force, N. This labor force is distributed between production activities, research activities, teaching activities and studying activities. We neglect any costs for migration and professional changes. Perfect competition is assumed to prevail in all markets.

We select commodity 1 produced by region 1 to serve as numeraire, with all the other prices being measured relative to its price. To simplify the analysis, we neglect the costs of transporting commodities between and within regions. It can be shown that we may introduce these transportation costs by taking Samuelson's "iceberg" form, in which transportation costs are incurred in the commodity transported. It should be noted that by adding transportation costs some of our results on geographical pattern formation might become stronger. However, as the model we present in this paper already is very complicated, we will assume zero transportation costs at this initial stage. The assumption of zero transportation costs of commodities implies price equality for both commodities between the two regions. As amenity factors are immobile, wage rates are not necessarily identical between the two regions. We assume, however, that wage rates are identical within each region, irrespective of occupation or profession.

We now introduce

i, s — subscript index for manufacturing industry and service industry, respectively

$F_{ij}(t)$ and $F_{sj}(t)$ — the output levels of region j's manufacturing and service sectors at time t, $j = 1, 2$;

$N_{ij}(t)$ and $N_{sj}(t)$ — the labor force employed by region j's manufacturing and service sectors at time t, $j = 1, 2$;

$N_r(t)$ — the number of teachers (and researchers) employed by the university;

$N_{ej}(t)$ — the number of students from region j studying at the university;

$N_j(t)$ — the number of people from the labor force living in region j;

$Z(t)$ — the knowledge stock of the whole economy;

$H_j(t)$ — region j's level of human capital;

$w_j(t)$ — region j's wage level;

$p(t)$ and $p_j(t)$ — the price of commodity 2 and the price of services in region j, and

$\tau(t)$ — the tax rate at time t.

Service production

We assume that service production is done by combining human capital and labor. The production function of the service sectors in the two regions are specified as follows

$$F_{sj} = a_{sj} H_j^{m_{sj}} N_{sj}, \quad a_{sj} > 0, \quad m_{sj} \geq 0, \quad j = 1, 2 \tag{11.1.1}$$

where a_{sj} and m_{sj} are parameters. We use the variable $H_j(t)$, to measure, in an aggregated sense, the average level of human capital of region j at time t. A similar variable has been widely used in the literature on human capital and economic growth. We only use labor and human capital as input factors in economic production. It should be noted that the introduction of other factors of production such as capital may cause great analytical difficulties.

The marginal conditions of service production are given by

$$w_j = \frac{(1-\tau) p_j F_{sj}}{N_{sj}}, \quad j = 1, 2. \tag{11.1.2}$$

We specify the production functions for the two regions' manufacturing sectors as follows

$$F_{ij} = a_{ij} H_j^{m_{ij}} N_{ij}, \quad a_{ij} > 0, \quad m_{ij} \geq 0, \quad j = 1, 2 \tag{11.1.3}$$

where a_{ij} and m_{ij} are parameters. The conditions for zero-profit are given by

$$w_1 = \frac{(1-\tau) F_{i1}}{N_{i1}}, \quad w_2 = \frac{(1-\tau) p F_{i2}}{N_{i2}}. \tag{11.1.4}$$

We assume that the utility level a person obtains in a given region is dependent on the region's amenity level, and on the consumption levels of services and commodities produced by manufacturing industry. We specify the utility functions as follows

$$U_j(t) = A_j \left(\frac{C_{sj}}{N_j} \right)^{\gamma} \left(\frac{C_{ij}}{N_j} \right)^{\zeta} \left(\frac{C_{2j}}{N_j} \right)^{\xi}, \quad j = 1, 2, \quad \gamma, \zeta, \xi > 0,$$

$$\gamma + \zeta + \xi = 1 \tag{11.1.5}$$

in which C_{sj}/N_j, C_{1j}/N_j and C_{2j}/N_j are the per capita consumption levels of services, commodity 1 and commodity 2, respectively, in region j at time t. In (11.1.5), A_j denotes region j's amenity level. The amenity level is an aggregated measure of regional living conditions, such as infrastructure conditions, cultural activities and climate in the two regions. Some locational amenities such as the pollution level, residential density and transportation congestion may depend upon the activities of the economic agents themselves. Other amenities, such as climate, transportation infrastructure and the existence of historical buildings and sites, are on the other hand only to a very limited extent affected by the activities of the economic agents, at least, in a short term perspective. Accordingly, in a strict sense, it is necessary to classify amenities into endogenous and exogenous amenities when modelling economic geography. Hence, which kinds of amenities that should be classified as endogenous or exogenous depend upon the time scale of the analysis and the characteristics of the economic system under consideration.

The assumption that the utility levels are identical over space at any point of time is represented by

$$U_1(t) = U_2(t). \tag{11.1.6}$$

The consumer problem is defined by

$$\max U_j(t)$$

subject to

$$p_j C_{sj} + C_{1j} + pC_{2j} = w_j N_j. \tag{11.1.7}$$

We then have the following optimal solutions

$$C_{sj} = \frac{\gamma w_j N_j}{p_j}, \quad C_{1j} = \zeta w_j N_j, \quad C_{2j} = \frac{\xi w_j N_j}{p}, \quad j = 1, 2. \tag{11.1.8}$$

The balances of demand and supply of commodities and services are given by

$$C_{j1} + C_{j2} = F_{ij}(t), \quad C_{sj} = F_{sj}(t), \quad j = 1, 2. \tag{11.1.9}$$

The assumption that the labor force is fully employed is represented by

$$N_{i1} + N_{s1} + N_{e1} + N_{e2} + N_r = N_1(t), \quad N_{i2} + N_{s2} = N_2(t),$$
$$N_1(t) + N_2(t) = N. \tag{11.1.10}$$

We assume for each region that the number of students is a fixed share of the number of workers. This is represented by

$$N_{ej} = n_e(N_{ij} + N_{sj}), \quad j = 1, 2 \tag{11.1.11}$$

where n_e is a positive parameter.

Teachers do teaching and research. Denote the total working time of each teacher by T. We assume that each teacher devotes the share t_r of his total working time to research. This implies that each teacher's teaching time is equal to $(1-t_r)T$ and his or her research time is equal to $t_r T$. We assume that the number of teachers is determined by the following equation

$$(1-t_r)TN_r = n_r N_e, \quad N_e \equiv N_{e1} + N_{e2} \tag{11.1.12}$$

where $N_{e1} + N_{e2}$ is the total number of students and $(1-t_r)TN_r$ is the total teaching time delivered by the teachers. The parameter, n_r, is equal to the teaching time devoted on average to each student.

We assume that the students and the teachers are paid by the tax income that the government collects and that they are paid at the same wage rate as the workers in region 1. Obviously, this assumption is strict, since we would expect instead that students may get a lower "wage" rate than workers in region 1 and that teachers may get a higher wage rate. Since the total tax income is used for paying the students and the teachers, we have the following balance equation

$$w_1(N_e + N_r) = \tau(F_{i1} + pF_{i2} + p_{s1}F_{s1} + p_{s2}F_{s2}). \tag{11.1.13}$$

Knowledge accumulation

In this chapter, we assume that only the university can create new knowledge. We neglect any other sources of knowledge creation such as learning by doing within the production sectors and knowledge creation through the pursuit of knowledge during leisure time. We specify the knowledge accumulation function as follows

$$\frac{dZ}{dt} = \tau_r Z^g (t_r T H_1 N_r)^h - \delta_z Z \tag{11.1.14}$$

in which δ_z is the fixed depreciation rate of knowledge. We call the term, $\tau_r Z^g (t_r T H_1 N_r)^h$, the contribution by the university to knowledge growth. This term can be interpreted in the following way. The knowledge production of the university is positively related to the total working time of scientists. To interpret the parameter, g, we note that on the one hand, as the knowledge stock is increased, the university may more effectively utilize existing knowledge to discover new theorems but that on the other hand, a large stock of existing knowledge may make the discovery of new knowledge more difficult. This implies that the parameter g may be either positive or negative, depending on the conditions for the knowledge creation which occurs at the university.

Accumulation of human capital

We assume that there are two general sources of improving a region's human capital, namely through education and through learning by producing. We propose that each region's human capital dynamics is given by

$$\frac{dH_j}{dt} = \tau_j Z^{g_j} (n_r H_j N_{ej})^{h_j} + \frac{\tau_{ij} Z^{g_{ij}} F_{ij}}{H_j^{\varepsilon_{ij}}} + \frac{\tau_{sj} Z^{g_{sj}} F_{sj}}{H_j^{\varepsilon_{sj}}} - \delta_h H_j \tag{11.1.15}$$

where $\delta_h (\geq 0)$ is the depreciation rate of human capital, and g_j, τ_j, τ_{ij}, τ_{sj}, and h_j ($j = 1, 2$), are non-negative parameters. The term $\tau_j Z^{g_j} (n_r H_j N_{ej})^{h_j}$ describes the contribution of education to human capital improvements in region j. The human capital of region j tends to increase with n_r, i.e. with the teaching time on average devoted to each student. We also assume that an increase in knowledge makes teaching more efficient, i.e., $g_j \geq 0$.

We only take account of learning by doing effects in human capital accumulation via the terms $\tau_{ij} Z^{g_{ij}} F_{ij} / H_j^{\varepsilon_{ij}}$ and $\tau_{sj} Z^{g_{sj}} F_{sj} / H_j^{\varepsilon_{sj}}$. The term, $\tau_{i1} Z^{g_{j1}} F_{i1} / H_j^{\varepsilon_{i1}}$, implies, for instance, that the contribution of region 1's manufacturing sector to the human capital improvement is positively related to its production scale, F_{i1}, and is dependent on the current level of the knowledge stock and the region's human capital level. The term, $H^{\varepsilon_{i1}}$, takes account of returns to scale effects in human capital accumulation. The case of $\varepsilon_{i1} > 0$ (< 0) implies that as human capital is increased it is more difficult (easier) to further improve the level of human capital.

We have now finished the model building. The system has 30 variables, $N_{sj}, N_{ij}, N_{ej}, H_j$, $F_{sj}, F_{ij}, C_{sj}, C_{1j}, C_{2j}$, U_j, N_j, w_j, p_j, ($j = 1, 2$), p, τ, N_r, and Z. It contains the same number of independent equations. We now continue with examining some properties of the dynamic system.

11.2 The Regional Equilibrium with Universities

This section is concerned with the conditions for the existence of equilibria and for stability. First, we show that for any given knowledge stock, $Z(t)$, and human capital, $H_1(t)$, and $H_2(t)$, the division of labor and the economic geography are uniquely determined at any point of time. The following proposition is proved in the Appendix.

Proposition 11.2.1
For any given levels of knowledge and human capital, $Z(t), H_1(t)$, and $H_2(t)$, all the other variables in the system are uniquely determined at any point of time. The values of the variables are given as functions of $H_1(t)$, and $H_2(t)$ by the following procedure: $w = w_1/w_2$ by (11.A.1.9) $\rightarrow \tau$ by (11.A.1.13) $\rightarrow N_j$, $j = 1, 2$, by (11.A.1.11) $\rightarrow N_{ij}$ by (11.A.1.6) $\rightarrow N_{sj} \rightarrow N_{ej}$ by (11.1.11) $\rightarrow N_r$ by (11.1.12) $\rightarrow F_{sj}$ by (11.1.1) $\rightarrow F_{ij}$ by (11.1.3) $\rightarrow w_1$ by (11.1.4) $\rightarrow w_2 = w_1/w \rightarrow p_j$ by (11.1.2) $\rightarrow p$ by (11.1.4) $\rightarrow C_{sj}, C_{ij}$ and C_{2j} by (11.1.8) $\rightarrow U_j$ by (11.1.5).

The above proposition guarantees that if we can find knowledge and human capital, $Z(t)$, $H_1(t)$, and $H_2(t)$, then we can explicitly solve all other variables as functions of $H_1(t)$ and $H_2(t)$ in the system at any point of time. This implies that if we find solutions of $Z(t)$ and $H_j(t)$ by the three differential equations (11.1.14) and (11.1.15), then we determine all the variables of the dynamic system. Hence, to examine the dynamic properties of the system, it is sufficient to examine the dynamic properties of (11.1.14) and (11.1.15). To simplify the discussion, we set up the following requirement.

Assumption 3.1
In the reminder of this chapter, we require $\tau_{sj} = \tau_{ij} = t_r = \delta_z = 0$ and $g_j = 0$ or $Z = 1$ to make sure $Z^{g_j} = 1$ in (11.1.15).

This requirement implies that we neglect human capital improvements through learning by doing and that we assume that knowledge is constant. If we don't make this omission, we have to discuss more cases of different combinations of parameters. As these discussions will provide little new insight, we limit our analysis to this special case. Under this requirement the dynamics of (11.1.14) and (11.1.15) are given by

$$\frac{dH_j}{dt} = \tau_j (n_r H_j N_{ej})^{h_j} - \delta_h H_j, \quad j = 1, 2. \tag{11.2.1}$$

In the above equations N_r is given by (11.A.1.5) and is independent of H_j, but N_{ej} are dependent on H_j. Equilibrium of the system is determined by

$$\tau_j (n_r H_j N_{ej})^{h_j} = \delta_h H_j, \quad j = 1, 2. \tag{11.2.2}$$

We may re-write the above equations for $j = 1$ and 2 separately, as follows

$$N_{e1} = a_1 H_1^a, \quad N_{e2} = a_2 H_2^b \tag{11.2.3}$$

where we require $h_j \neq 1$, $a \equiv (1 - h_1)/h_1$, $b \equiv (1 - h_2)/h_2$ and

$$a_j \equiv \frac{(\delta_h / \tau_j)^{1/h_j}}{n_r}, \quad j = 1, 2.$$

Since $h_j > (<) 1$, we see that a and b may be either negative or positive. By using (11.2.3) and $N_{e1} + N_{e2} = n_e N / n^*$ in (11.A.1.5), we solve

$$H_2 = \varphi(H_1) \equiv \left(\frac{n_e N / n^* - a_1 H_1^a}{a_2} \right)^{1/b} \tag{11.2.4}$$

where

$$n^* \equiv 1 + n_e n_0, \quad n_0 \equiv 1 + \frac{n_r}{(1 - n_r) T}.$$

We see that H_2 can be expressed as a unique function of H_1. By using $w = AF$ in (11.A.1.10) and (11.2.4), we get

$$w = \phi(H_1) \equiv \frac{a_0 A \varphi^{m_2}(H_1)}{H_1^{m_1}} \tag{11.2.5}$$

where $m_j = m_{sj}\gamma/(1-\gamma) > 0$. By using $N_{e2} = a_2 H_2^b$ in (11.2.3) and $N_{e2} = n_e N_2$ in (11.1.10) and (11.1.11), we get $a_2 H_2^b = n_e N_2$. Substituting (11.2.4), (11.2.5) and N_2 in (11.A.1.12) into this equation yields

$$\Phi(H_1) \equiv a^* \varphi^b - \Phi_0(H_1) = 0 \tag{11.2.6}$$

where $a^* \equiv a_2 n^*/n_e N > 0$, and

$$\Phi_0 \equiv \frac{\phi\Phi^*(H_1)}{(1-\tau)(\phi-1)(\zeta+\xi\phi)},$$
$$\Phi^*(H_1) \equiv \xi n^* \phi(1-\tau) + n^*(\gamma+\zeta)(1-\tau) - 1.$$

It can be seen that for any solution H_1 of (11.2.6) which satisfies $\phi(H_1) > 1$ and $0 < H_1 < \infty$, the system has a unique equilibrium. In other words, the number of economic equilibria is equal to the number of meaningful solutions of (11.2.6).

In order to guarantee the existence of solutions, we examine the case of $a > 0$ and $b > 0$. In this case, given that $H_2 > 0$ it is necessary to require: $0 < H_1 < H_0 (\equiv (n_e N/a_1 n^*)^{1/a}$ at which $\varphi(H_0) = 0$. At $H_1 = 0$ and $\varphi > 0$: $\phi \to \infty$. One can directly check that in the interval $(0, H_0)$ there is a unique H^* at which $\phi = 1$. We now show that $\Phi(H_1) = 0$ has at least one solution for $0 < H_1 < H^*$. We have: $\Phi(0) > 0$ and $\Phi(H^*) < 0$. Hence, $\Phi(H_1) = 0$ has at least one solution for $0 < H_1 < H^*$ with $w > 1$ at this solution. Taking derivatives of Φ with respect to H_1 yields

$$\Phi' = a^* b \varphi^{b-1} \varphi' + \Phi_0 \left\{ \frac{(\xi\phi + \gamma + \zeta)\tau' - \xi\phi'(1-\tau)}{\Phi^*} n^* \right.$$

$$\left. - \frac{\tau}{1-\tau} + \frac{\phi''}{\phi(\phi-1)} + \frac{\xi\phi'}{\zeta + \xi\phi} \right\}$$

(11.2.7)

in which

$$\varphi' = - \frac{a_1 a H_1^{a-1} \varphi}{b(n_e N / n^* - a_1 H_1^a)} < 0, \quad \phi' = \frac{m_2 \varphi' \phi}{\varphi} - \frac{m_1 \phi}{H_1} < 0,$$

$$\tau' = \frac{(1-\tau)(n^* - 1)\xi\phi'}{2\gamma n^* (\tau + W_1)} < 0 \tag{11.2.8}$$

where τ is given by (11.A.1.13) and $\tau + W_1 > 0$. Since the sign of Φ' is ambiguous in the interval $(0, H^*)$, it is difficult to explicitly determine the uniqueness of a solution. If $\Phi' < 0$ holds in the interval, then the system has a unique equilibrium. By (11.2.7) we see that this requirement is satisfied if, for instance, the terms $(\xi\phi + \gamma + \zeta)\tau' - \xi\phi'(1-\tau)$ and

$$- \frac{\tau'}{1-\tau} + \frac{\xi\phi'}{\zeta + \xi\phi}$$

have the same sign as that of $\phi' \, (< 0)$.

In the case of $a < 0$ and $b < 0$, to guarantee that $H_2 > 0$ it is necessary to require: $0 > H_1 > H_0$. At $H_1 = H_0$, $\varphi \to \infty$ and $\phi \to \infty$. One can directly check that in the interval (H_0, H^*), there is a unique H^* at which $\phi = 1$. We now show that $\Phi(H_1) = 0$ has at least one solution for $H_0 < H_1 < H^*$. We have: $\Phi(H_0) > 0$ and $\Phi(H^*) < 0$. Hence, $\Phi(H_1) = 0$ has at least one solution $H_0 < H_1 < H^*$ with $w > 1$ at this solution. Similarly, we may discuss cases of $a > 0$ and $b < 0$ or $a < 0$ and $b > 0$. From the above discussion, we have the following proposition.

Proposition 11.2.2.
We are concerned with the economic dynamics under Assumption 3.1. We further assume $a > 0$ and $b > 0$ (i.e., $h_j < 1$). The system has at least one equilibrium. In particular, if Φ' is given by (11.2.7) is negative for $0 < H_1 < H^*$, then the equilibrium is unique. The equilibrium values of H_1 and H_2 are given by (11.2.6) and (11.2.4), respectively. The equilibrium values of the other variables are given by the procedure given by Proposition 11.2.1.

The assumption, $h_j < 1$, means that human capital improvement exhibits a decreasing returns to scale effect. Here, we don't provide the conditions for stability, since the calculation of eigenvalues is too complicated. In the reminder of this chapter, we will examine the impact on the economic geography of changes in some parameters.

11.3 The Education Policy and the Regional Structure

This section is concerned with the impact of the education policy parameter, n_e, on the economic geography. Taking the derivatives of (11.A.1.5) with respect to n_e, we get

$$\frac{1}{N_e}\frac{dN_e}{dn_e} = \frac{1}{N_r}\frac{dN_r}{dn_e} = \frac{1}{n_e n^*} > 0. \tag{11.3.1}$$

As the ratio, n_e, between the number of students and the total number of workers in the economy engaged in manufacturing and service production is increased, the number of students and the number of teachers also increases. Taking the derivatives of (11.2.6) with respect to n_e yields

$$-\Phi'\frac{dH_1}{dn_e} = ba^*\varphi^{b-1}\frac{d\varphi}{dn_e} + \Phi_0\left\{\left(\frac{\xi}{\zeta + \xi\phi} + \frac{1}{\phi - 1}\right)\frac{d\phi}{dn_e}\right.$$
$$\left. - \frac{1}{1-\tau}\frac{d\tau}{dn_e}\right\} - \left[\xi n_e(1-\tau)\frac{d\phi}{dn_e} - (\xi\phi + \gamma + \zeta)n^*\frac{d\tau}{dn_e}\right.$$
$$\left. + (\xi\phi + \gamma + \zeta)(1-\tau)n_0\right]\frac{\Phi_0}{\Phi^*} \tag{11.3.2}$$

in which we use

$$\frac{d\tau}{dn_e} = \frac{n_0(\zeta + \xi\phi)(1-\tau) + (n^* - 1)(1-\tau)n^* \xi d\phi/dn_e + \tau n_0}{(\tau + W_1)2\gamma n^{*2}},$$

$$\frac{1}{\varphi}\frac{d\varphi}{dn_e} = \frac{1}{m_2\phi}\frac{d\phi}{dn_e} = \frac{n_e N/n^* - a_1 H_1^a}{bn^{*2}} N > 0. \tag{11.3.3}$$

Since the terms

$$\left(\frac{\xi}{\zeta + \xi\phi} + \frac{1}{\phi - 1}\right)\frac{d\phi}{dn_e} - \frac{1}{1-\tau}\frac{d\tau}{dn_e}$$

and

$$\xi n_e(1-\tau)\frac{d\phi}{dn_e} - (\xi\phi + \gamma + \zeta)n^* \frac{d\tau}{dn_e} + (\xi\phi + \gamma + \zeta)(1-\tau)n_0$$

may be either positive or negative, an increase in the number of students may either increase or decrease the human capital level in region 1. By using (11.2.4) and (11.2.5) we obtain

$$\frac{dH_2}{dn_e} = \varphi'\frac{dH_1}{dn_e} + \frac{d\varphi}{dn_e}, \quad \frac{dw}{dn_e} = \phi'\frac{dH_1}{dn_e} + \frac{d\phi}{dn_e},$$

$$\frac{d\tau}{dn_e} = \tau'\frac{dH_1}{dn_e} + \frac{d\tau}{dn_e} \tag{11.3.4}$$

where $\varphi'(<0)$, $\phi'(<0)$ and $\tau'(<0)$ are given by (11.2.8). In the case $dH_1/dn_e < 0$, we have: $dH_2/dn_e > 0$, $dw/dn_e > 0$ and $d\tau/dn_e > 0$. This means that if an increase in the number of students reduces the human capital of region 1, then as a result of that decrease, the human capital of region 2, H_2, is increased, the wage difference, $w(>1)$, between the two regions is enlarged, and the tax rate is increased. But if an increase in the number of students increases the human capital of region 1, the impact on region 2's human capital, the impact on the wage difference between the two regions and the impact on the tax rate are ambiguous.

By using (11.2.3) we get

$$\frac{dN_{e1}}{dn_e} = \frac{aN_{e1}}{H_1}\frac{dH_1}{dn_e}, \quad \frac{dN_{e2}}{dn_e} = \frac{bN_{e2}}{H_2}\frac{dH_2}{dn_e}. \tag{11.3.5}$$

We see that if dH_1/dn_e is negative (positive), then as the number of students increases, the number of region 1's students is reduced (increased); the number of region 2's students is increased (reduced). From $N_2 = N_{e2}/n_e$ and $N_1 = N - N_2$, we get the impact on the distribution of the total labor force between the two regions as follows

$$\frac{dN_1}{dn_e} = -\frac{dN_2}{dn_e} = \frac{N_2}{n_e} - \frac{bN_2}{H_2}\frac{dH_2}{dn_e}. \tag{11.3.6}$$

If $dH_2/dn_e < 0$, then some people migrate from region 2 to region 1 as the number of students is increased. If $dH_2/dn_e > 0$, the impact on the distribution of the total labor force between the two regions is ambiguous. From (11.A.1.6) and

$$N_{i2} = N_2 - N_{s2}, \quad N_{1i} = N_1 - N_{s1} - \left(1 - \frac{1}{n^*}\right)N,$$

we get

$$\frac{dN_{sj}}{dn_e} = -\gamma N_j\frac{d\tau}{dn_e} + (1-\tau)\gamma\frac{dN_j}{dn_e},$$
$$\frac{dN_{i1}}{dn_e} = (1 - \gamma + \gamma\tau)\frac{dN_1}{dn_e} + \gamma N_1\frac{d\tau}{dn_e} + \frac{n_0 N}{n^{*2}},$$
$$\frac{dN_{i2}}{dn_e} = (1 - \gamma + \gamma\tau)\frac{dN_2}{dn_e} + \gamma N_2\frac{d\tau}{dn_e}. \tag{11.3.7}$$

From (11.1.2), (11.1.4) and $w_2 = w_1/w$, we get

$$\frac{1}{w_1}\frac{dw_1}{dn_e} = -\frac{1}{1-\tau}\frac{d\tau}{dn_e} + \frac{m_{i1}}{H_1}\frac{dH_1}{dn_e},$$
$$\frac{1}{w_2}\frac{dw_2}{dn_e} = \frac{1}{w_1}\frac{dw_1}{dn_e} - \frac{1}{w}\frac{dw}{dn_e},$$

$$\frac{1}{p}\frac{dp}{dn_e} = \frac{1}{w_2}\frac{dw_2}{dn_e} + \frac{1}{1-\tau}\frac{d\tau}{dn_e} - \frac{m_{i2}}{H_2}\frac{dH_2}{dn_e},$$
$$\frac{1}{p_j}\frac{dp_j}{dn_e} = \frac{1}{w_j}\frac{dw_j}{dn_e} + \frac{1}{1-\tau}\frac{d\tau}{dn_e} - \frac{m_{sj}}{H_j}\frac{dH_j}{dn_e}. \tag{11.3.8}$$

Taking the derivatives of (11.1.1), (11.1.3) and (11.1.8) with respect to n_e, we directly have

$$\frac{1}{F_{sj}}\frac{dF_{sj}}{dn_e} = \frac{1}{C_{sj}}\frac{dC_{sj}}{dn_e} = \frac{m_{sj}}{H_j}\frac{dH_j}{dn_e} + \frac{1}{N_{sj}}\frac{N_{sj}}{dn_e},$$
$$\frac{1}{F_{ij}}\frac{dF_{ij}}{dn_e} = \frac{m_{ij}}{H_j}\frac{dH_j}{dn_e} + \frac{1}{N_{ij}}\frac{N_{ij}}{dn_e},$$
$$\frac{1}{C_{ij}}\frac{dC_{ij}}{dn_e} = \frac{1}{w_j}\frac{dH_j}{dn_e} + \frac{1}{N_j}\frac{N_j}{dn_e}. \tag{11.3.9}$$

11.4 The Amenity Level and the Regional Structure

This section is concerned with the impact of changes in regional amenities, A_1, and A_2, on the economic geography. From $A = (A_2 / A_1)^{1/(1-\gamma)}$, we see that an increase in A may be caused by an increase in A_2 or a decrease in A_1. One can see that the equilibrium values of all the variables, except U_j, are affected by the ratio between A_2 and A_1. Hence, it is sufficient to take the derivatives of the variables with respect to A. The derivatives of (11.A.1.5) with respect to A, we get

$$\frac{dN_e}{dA} = \frac{dN_r}{dA} = 0. \tag{11.4.1}$$

When the amenity level of region 2 is changed, the number of students, N_e, and the number of teachers, N_r are not affected.

Taking the derivatives of (11.2.6) with respect to A yields

$$\Phi'\frac{dH_1}{dA} = \left\{\xi n^*(1-\tau) - \left[\frac{\xi\Phi^*}{\zeta + \xi\phi} + \frac{\Phi^*}{\phi - 1}\right]\frac{\phi}{A} - \frac{1}{1-\tau}\frac{d\tau}{dA}\right\}\frac{\Phi_0}{\Phi^*} \tag{11.4.2}$$

where we use

$$\frac{d\varphi}{dA} = 0, \quad \frac{1}{\phi}\frac{d\phi}{dA} = \frac{1}{A}, \quad \frac{d\tau}{dA} = \frac{(n^* - 1)(1-\tau)\phi}{2\gamma n^* A(\tau + W_1)} > 0. \tag{11.4.3}$$

Since the terms Φ' and $\xi n^*(1-\tau)\phi/A - [1/(1-\tau)]d\tau/dA$ may be either positive or negative, we see that an increase in the amenity level of region 2 may either increase or decrease the human capital level of region 1. From (11.2.4) and (11.2.5), we obtain

$$\frac{dH_2}{dA} = \varphi'\frac{dH_1}{dA}, \quad \frac{dw}{dA} = \phi'\frac{dH_1}{dA} + \frac{\phi}{A}, \quad \frac{d\tau}{dA} = \tau'\frac{dH_1}{dA} + \frac{d\tau}{dA} \tag{11.4.4}$$

where we use (11.4.2) and $\varphi'(<0)$, $\phi'(<0)$ and $\tau'(<0)$ are given by (11.2.8). In the case of $dH_1/dA < 0$, $dH_2/dA > 0$, $dw/dA > 0$ and $d\tau/dA > 0$. This means that if an increase in the amenity level of region 2 reduces the human capital of region 1, then as a result of this increase, region 2's human capital, H_2, is improved, the wage difference, $w\,(>1)$, between the two regions is enlarged, and the tax rate is increased. But if an increase in the amenity level of region 2 increases the human capital of region 1, then the impact on the human capital of region 2 is negative but the impact on the wage difference between the two regions and on the tax rate are ambiguous.

By using (11.2.3), we get

$$\frac{dN_{e1}}{dA} = \frac{aN_{e1}}{H_1}\frac{dH_1}{dA}, \quad \frac{dN_{e2}}{dA} = \frac{bN_{e2}}{H_2}\frac{dH_2}{dA}. \tag{11.4.5}$$

We see that if dH_1/dA is negative (positive), then as a result of increasing the amenity level of region 2, the number of region 1 students is reduced (increased) and the number of region 2 students is increased (reduced). From

$N_2 = N_{e2}/n_e$ and $N_1 = N - N_2$, we get the impact on the distribution of the total labor force between the two regions as follows

$$\frac{dN_1}{dA} = -\frac{dN_2}{dA} = -\frac{bN_2}{H_2}\frac{dH_2}{dA}. \tag{11.4.6}$$

If improved living conditions in region 2 reduces the human capital of region 1, then as a result of the improved conditions, some people of region 1 will migrate to region 2 and vice versa. By using (11.A.1.6), $N_{i2} = N_2 - N_{s2}$ and $N_{1i} = N_1 - N_{s1} - (1 - 1/n^*)N$, we get

$$\frac{dN_{sj}}{dA} = -\gamma N_j \frac{d\tau}{dA} + (1-\tau)\gamma \frac{dN_j}{dA},$$
$$\frac{dN_{ij}}{dA} = (1 - \gamma + \gamma\tau)\frac{dN_j}{dA} + \gamma N_j \frac{d\tau}{dA}, \quad j = 1, 2. \tag{11.4.7}$$

By using (11.1.2), (11.1.4) and $w_2 = w_1/w$, we get

$$\frac{1}{w_1}\frac{dw_1}{dA} = -\frac{1}{1-\tau}\frac{d\tau}{dA} + \frac{m_{i1}}{H_1}\frac{dH_1}{dA}, \quad \frac{1}{w_2}\frac{dw_2}{dA} = \frac{1}{w_1}\frac{dw_1}{dA} - \frac{1}{w}\frac{dw}{dA},$$
$$\frac{1}{p}\frac{dp}{dA} = \frac{1}{w_2}\frac{dw_2}{dA} + \frac{1}{1-\tau}\frac{d\tau}{dA} - \frac{m_{i2}}{H_2}\frac{dH_2}{dA},$$
$$\frac{1}{p_j}\frac{dp_j}{dA} = \frac{1}{w_j}\frac{dw_j}{dA} + \frac{1}{1-\tau}\frac{d\tau}{dA} - \frac{m_{sj}}{H_j}\frac{dH_j}{dA}. \tag{11.4.8}$$

Taking the derivatives of (11.1.1), (11.1.3) and (11.1.8) with respect to A, we directly have

$$\frac{1}{F_{sj}}\frac{dF_{sj}}{dA} = \frac{1}{C_{sj}}\frac{dC_{sj}}{dA} = \frac{m_{sj}}{H_j}\frac{dH_j}{dA} + \frac{1}{N_{sj}}\frac{N_{sj}}{dA},$$
$$\frac{1}{F_{ij}}\frac{dF_{ij}}{dA} = \frac{m_{ij}}{H_j}\frac{dH_j}{dA} + \frac{1}{N_{ij}}\frac{N_{ij}}{dA},$$
$$\frac{1}{C_{Ij}}\frac{dC_{Ij}}{dA} = \frac{1}{w_j}\frac{dH_j}{dA} + \frac{1}{N_j}\frac{N_j}{dA}. \tag{11.4.9}$$

11.5 On Regional Dynamics with Universities

This chapter proposed a dynamic two-region model with endogenous knowledge and human capital accumulation that describes the dynamic interdependence between knowledge creation, human capital accumulation, the regional division of labor, the spatial price structure under perfect competition and government intervention in R&D and higher education. First, we provided conditions for the existence of equilibria. Second, we examined the impacts of a change in the education policy parameter - the share of the labor force participating in higher education - on the economic geography. Third we examined the impacts of changes in regional amenities on the economic geography.

Appendix

A.11.1 Proving Proposition 11.2.1

Substituting (11.1.11) into (11.1.12) yields

$$N_r = \frac{N_{i1} + N_{s1} + N_{i2} + N_{s2}}{(1 - t_r)T} n_e n_r . \tag{11.A.1.1}$$

By using (11.1.10) and (11.1.12), we have

$$N_{i1} + N_{si} + (N_{e1} + N_{e2})\left\{1 + \frac{n_r}{(1 - t_r)T}\right\} = N_1 .$$

Substituting $N_{ej} = n_e (N_{ij} + N_{sj})$ into the equation above yields

$$n^* (N_{i1} + N_{s1}) + (n^* - 1)(N_{i2} + N_{s2}) = N_1 \tag{11.A.1.2}$$

where $n^* \equiv 1 + n_e n_0$ and $n_0 \equiv 1 + n_r / T(1 - t_r)$. By using (11.A.1.2), $N_1 + N_2 = N$ and $N_2 = N_{i2} + N_{s2}$, we get

$$N_{i1} + N_{s1} + N_{i2} + N_{s2} = \frac{N}{n^*} . \tag{11.A.1.3}$$

From (11.A.1.3) and (11.1.10), we get

$$N_{e1} + N_{e2} + N_r = \frac{n^* - 1}{n^*} N. \tag{11.A.1.4}$$

By using (11.A.1.4) and (11.1.12), we have

$$N_r = \frac{n_e n_r N}{(1 - t_r) n^* T}, \quad N_{e1} + N_{e2} = \frac{n_e N}{n^*}. \tag{11.A.1.5}$$

From (11.1.8) and (11.1.2), we get $C_{sj} = (1 - \tau)\gamma N_j F_{sj} / N_{sj}$. From this equation and $C_{sj} = F_{sj}$, we get

$$N_{sj} = (1 - \tau)\gamma N_j. \tag{11.A.1.6}$$

Substituting w_j in (11.1.4), $C_{1j} = \zeta w_j N_j$ and $C_{2j} = \xi w_j N_j / p$ in (11.1.8) into $C_{j1} + C_{j2} = F_{ij}$, we get

$$N_{i1} = \zeta(1 - \tau)\left(N_1 + \frac{N_2}{w}\right), \quad N_{i2} = \xi(1 - \tau)(wN_1 + N_2) \tag{11.A.1.7}$$

where we use (11.1.3) and $w \equiv w_1 / w_2$. By using (11.1.2), we get

$$w = \frac{a_{s1} p_1 H_1^{m_{s1}}}{a_{s2} p_2 H_2^{m_{s2}}}. \tag{11.A.1.8}$$

Substituting (11.1.8) into (11.1.5) and then using (11.1.6), we get

$$w = \frac{A_2}{A_1}\left(\frac{p_1}{p_2}\right)^{\gamma}. \tag{11.A.1.9}$$

By using (11.A.1.8) and (11.A.1.9), we solve

$$\frac{p_1}{p_2} = AF^{1/\gamma}, \quad w = AF \tag{11.A.1.10}$$

where

$$A \equiv \left(\frac{A_2}{A_1}\right)^{1/(1-\gamma)}, \quad F(H_1, H_2) \equiv \frac{a_0 H_2^{m_2}}{H_1^{m_1}}, \quad a_0 \equiv \left(\frac{a_{s2}}{a_{s1}}\right)^{\gamma/(1-\gamma)},$$

$$m_j \equiv \frac{m_{sj}\gamma}{1-\gamma}.$$

We see that p_1/p_2 and w can be expressed as functions of H_1 and H_2. Substituting (11.A.1.6) and (11.A.1.7) into (11.A.1.3) yields

$$wN_1 + N_2 = \frac{1 - \gamma n^*(1-\tau)}{(1-\tau)(\zeta + \xi w)n^*} wN. \tag{11.A.1.11}$$

From this equation and given that $N_1 + N_2 = N$, we solve

$$N_1 = \frac{w - \zeta n^*(1-\tau) - n^* w(\gamma + \xi)(1-\tau)}{(1-\tau)(w-1)(\zeta + \xi w)n^*} N,$$

$$N_2 = \frac{\xi n^* w(1-\tau) + n^*(\gamma + \zeta)(1-\tau) - 1}{(1-\tau)(w-1)(\zeta + \xi w)n^*} wN. \tag{11.A.1.12}$$

It is necessary to require $0 < N_1 < N$. Here we neglect the case of $w = 1$. In the case of $w > (<)1$, it is possible to directly check that the requirement is satisfied if

$$w > (<)\, n^*(1-\tau)[\zeta - w(\gamma + \xi)], \quad 1 < (>) n^*(1-\tau)(\gamma + \zeta + \xi w).$$

We are only concerned with the case when $w > 1$ at any point of time. If $\zeta \geq \gamma + \xi$, then $w > n^*(1-\tau)[\zeta - w(\gamma + \xi)]$ holds for the case $w > 1$. When explicitly solving τ, we show that $n^*(1-\tau)(\gamma + \zeta + \xi w) > 1$ in the case of $w > 1$. This implies that in the case of $w > 1$, we have a meaningful solution of both N_1 and N_2.

It is straightforward to show that

$$F_{i1} + pF_{i2} + p_{si}F_{si} + p_{s2}F_{s2} = w_1 N_1 + w_2 N_2.$$

Substituting this equation and (11.A.1.4) into (11.1.13) we get

$$\frac{n^* - 1}{n^*} wN = \tau(wN_1 + N_2).$$

By using the equation above and (11.A.1.10), we get

$$\tau^2 + 2W_1\tau - W_2 = 0, \tag{11.A.1.13}$$

where

$$W_1(H_1, H_2) \equiv \frac{(n^* - 1)(\zeta + \xi w) + 1 - \gamma n^*}{2\gamma n^*},$$

$$W_2(H_1, H_2) = \frac{(n^* - 1)(\zeta + \xi w)}{\gamma n^*} > 0.$$

We require $W_1 > 0$. It is straightforward to check that the sign of W_1 is the same as that of $\zeta + \xi + n_e n_0(\xi w + \zeta - \gamma)$, where $n_0 \equiv 1 + n_r / T(1 - t_r)$. We see that the requirement is satisfied if the term $(\xi w + \zeta - \gamma)$ is properly small. For any given $H_1(t)$ and $H_2(t)$, the equation (11.A.1.13) has two solutions, one solution is positive and the other is negative. The positive solution, $\tau(H_1, H_2)$, is given as

$$\tau(H_1, H_2) = (W_1^2 + W_2)^{1/2} - W_1. \tag{11.A.1.14}$$

According to the definition, we should require $\tau < 1$. This is guaranteed if $W_2 < 2W_1 + 1$. This is always satisfied given the definitions of W_1 and W_2. Hence, for any given H_1 and H_2 the solution given by (11.A.1.14) determines the tax rate, $0 < \tau < 1$. It is straightforward to show that this solution guarantees $1 < n^*(1 - \tau)(\gamma + \zeta + \xi w)$ in the case of $w > 1$, using $(\gamma + \zeta + \xi w) > 1$. This implies that the solution N_1 given by (11.A.1.12) satisfies $0 < N_1 < N$.

Proposition 11.2.1 shows how all the other variables might be expressed as functions of $H_1(t)$, $H_2(t)$ and $Z(t)$ at any point of time.

12 Regional Dynamics in an Isolated State

Chapter 6 propounded a regional model with spatial structures. But the model is static as we neglect dynamics of any growth factors. This chapter will extend the model proposed in Chapter 6 to deal with issues of regional dynamics with spatial structures. This chapter develops a dynamic urban model synthesizing the two main approaches in urban economics. The former is mainly initiated in the seminal work of von Thünen (1826) and further developed in the pioneering works by Isard (1956), Beckmann (1957), Wingo (1961) and Alonso (1964). The review of the current literature can be found, e.g., in Fujita (1989). This approach explains static spatial pattern without explaining growth mechanisms, such as capital and knowledge accumulation (not to say population growth and trades), of cities. The latter, i.e., the neoclassical urban growth theory, is mainly concerned with economic mechanisms of urban growth (e.g., Henderson, 1985). But it neglects structural pattern formation process. The purpose of this chapter is to suggest a compact framework by combining the basic ideas of these approaches to explain urban dynamics with endogenous urban pattern and capital and knowledge accumulation.

Despite widespread agreement that one of the important characteristics of spatial economy is the presence of scale effects in economic activities arising from the concentration of economic activities at a physically constrained site, only a few spatial models have been suggested to explain urban dynamics with scale effects of infrastructures, environment and population (e.g., Calem and Carlino, 1991, David and Rosenbloom, 1990, Anas, 1992, Helsley and Starnge, 1991, Helsley and Sullivan, 1991). Still less spatial economic models explicitly treat knowledge accumulation as an endogenous process. But as shown in the modern literature of economic growth and development, knowledge is extremely important in explaining scale and scope economies. This chapter tries to introduce endogenous knowledge to location theory to examine how economic spatial structure may be affected by knowledge. For simplicity of analysis, we are concerned with dynamics of an isolated economic system consisting of three - agricultural, industrial and service - kinds of economic activities. The geography of our economic system consists of the three parts, i.e., the CBD, the residential area and the agricultural area. Different economic activities are located spatially differently. We assume that industrial production and service supply are located at the CBD, urban households are located in the residential area (as in the Alonso model), and farming and farmers' dwelling location are distributed over the rural area. The model determines spatial and temporal distribution of real variables and monetary variables under the assumption of perfect competition in all the markets.

This chapter may be considered as a combination of the models proposed in Chapters 6 and 11. This chapter is organized as follows. Section 12.1 defines the dynamic urban model with endogenous land boundary and capital and knowledge accumulation. Section 12.2 guarantees the existence of temporary urban patterns for given capital and knowledge. Section 12.3 guarantees the existence of long-run urban equilibria and provides the stability conditions of the system. Section 12.4 examines the effects of changes in the labor force upon the urban structure. Section 12.5 concludes the chapter. The appendix provides the proofs of the results in Sections 12.2 and 12.3. This chapter is based on Zhang (1993c).

12.1 The General Model of the Isolated State

We consider an economic system consisting of three, industrial, service, and agricultural, sectors. The system geographically consists of the CBD, the residential area and the agricultural area. Let the form of the system is of cycle with a fixed radius. The system is monocentric; that is, it has a single pre-specified center of fixed size called the CBD (central business district). We assume that the industrial and service sectors are located at the CBD. The residential area is located outside the CBD and occupied by the households working in the CBD. The agricultural area lies outside the residential area. It is assumed that farmers choose dwelling sites where they do farming.

We select agricultural good to serve as numeraire. We assume that the wage rates are identical among different professions due to perfect competition in labor market. We assume that capital is freely mobile among the three sectors by perfectly competitive mechanism. This implies that the interest rate is identical among the three sectors. We introduce

N, L^* — the fixed total labor force and the radius of the system, respectively;

$L(t)$ — the distance from the CBD to the boundary between the residential area and the agricultural area;

ω — the distance from the CBD to a point either in the residential area or in the agricultural area;

subscripts, a, i and s — denoting the agricultural, industrial and the service sector, respectively;

$K(t)$ — the total capital stock in the system;

$N_j(t)$ and $K_j(t)$, $j = a, i, s$ — labor force and capital stock employed by sector j, respectively;

$F_j(t)$ and $C_j(t)$, $j = a, i, s$ — the total output of sector j and the total consumption of sector j 's product by the population, respectively;

$p_s(t)$, $p_i(t)$, $w(t)$, and $r(t)$ — the price of services, the price of industrial good, the wage rate, and the interest rate, respectively;

$R(\omega,t)$ — the land rent at location ω.

We now describe production of the three sectors and behavior of consumers.

The agricultural sector

Agricultural production is a process of combining land, labor force, capital, and knowledge. We propose the following production function of the agricultural sector

$$F_a(t) = Z^{m_a} K_a^{\alpha_a} N_a^{\beta_a} L_a^{\varsigma}, \quad m_a \geq 0, \ \alpha_a, \ \beta_a, \ \varsigma > 0, \quad \alpha_a + \beta_a + \varsigma = 1 \tag{12.1.1}$$

in which $Z(t)$ is the knowledge stock of the system and L_a is the land used by the agricultural sector. The marginal conditions are given by

$$r = \frac{\alpha_a F_a}{K_a}, \quad w = \frac{\beta_a F_a}{N_a}, \quad R_a = \frac{\varsigma F_a}{L_a}. \tag{12.1.2}$$

Assuming that agricultural good is not saved, we have

$$C_a = F_a. \tag{12.1.3}$$

The service sector

We propose the following production function of the service sector

$$F_s(t) = Z^{m_s} K_s^{\alpha_s} N_s^{\beta_s}, \quad m_s \geq 0, \ \alpha_s, \ \beta_s > 0, \ \alpha_s + \beta_s = 1. \tag{12.1.4}$$

The profit-maximization yields

$$r = \frac{\alpha_s p_s F_s}{K_s}, \quad w = \frac{\beta_s p_s F_s}{N_s}. \tag{12.1.5}$$

As services are consumed simultaneously as they are produced, we have

$$C_s = F_s. \tag{12.1.6}$$

The industrial sector
We specify the industrial production function in the following form

$$F_i(t) = Z^{m_i} K_i^{\alpha_i} N_i^{\beta_i}, \quad m_i \geq 0, \quad \alpha_i, \quad \beta_i > 0, \quad \alpha_i + \beta_i = 1. \quad (12.1.7)$$

The marginal conditions are given by

$$r = \frac{\alpha_i p_i F_i}{K_i}, \quad w = \frac{\beta_i p_i F_i}{N_i}. \quad (12.1.8)$$

Consumers' behavior
It is assumed that the revenue from land is equally shared among the population. The consumption budget of each individual is given by

$$y(t) = (1 - s)\frac{Y(t)}{N} \quad (12.1.9)$$

in which s is the given savings rate and Y is the gross income given by

$$Y = F_a + p_i F_i + p_s F_s. \quad (12.1.10)$$

In a strict sense, the gross income under the assumption of the public ownership should be written as

$$Y = rK + wN + 2\pi \int_0^{\omega} \omega R(\omega) d\omega = F_a + p_i F_i + p_s F_s +$$

the total land rent income from the housing market.

We omit the part of the total land rent from housing from the gross income defined in this chapter. One may explain this by assuming that the total land rent from housing market is distributed to public services, such as maintaining military forces. As shown later, this assumption analytically simplifies our study. Hence, we accept it for simplicity of analysis.

We assume that the utility of an individual from consuming industrial good, services and housing can be expressed in the following form

$$U(\omega,t) = c_a^{\mu}(\omega,t) c_s^{\gamma}(\omega,t) c_i^{\xi}(\omega,t) c_h^{\eta}(\omega,t), \quad \mu, \gamma, \xi, \eta > 0 \quad (12.1.11)$$

where $c_a(\omega,t)$, $c_s(\omega,t)$, $c_i(\omega,t)$, and $c_h(\omega,t)$ are consumption levels of agricultural good, services, industrial good and housing of a household at location ω and at time t, μ, γ, ξ and η are parameters. For simplicity of analysis, we specify values of μ, γ, ξ and η by $\mu + \gamma + \xi = \eta$. Although this hypothesis will not essentially affect our analytical conclusion, as shown late it helps us to simplify the analysis. Here, we neglect the possibility that profession choice may have effects upon utility levels. We also omit introducing leisure time into the utility function. We also neglect some kinds of externalities which may affect the form of utility function.

Although we assume that urban residents and farmers have the identical utility function and the same budget constraint, their consumption components may be different because of differences of geographical locations in dwelling sites and working places. As they have to travel to the CBD for working, urban residents have transportation cost burden which farmers don't have. We assume that it costs urban residents nothing to go to the CBD to consume product of the service sector (as they work there). Indeed, the assumption is not strictly true when we introduce family structures and holiday behavior into our system. On the other hand, as the service sector is located at the CBD, farmers have to go there in order to enjoy what city's service sector provides. Hence, it costs farmers extra transportation cost (in comparison to urban residents) in order to consume product of the service sector.

The consumer problem of an urban resident is defined by

$$Max\ U(c_a, c_s, c_i, c_h)$$

subject to

$$c_a + p_s c_s + p_i c_i + R c_h = y_c, \quad 0 \le \omega < L \tag{12.1.12}$$

where

$$y_c(\omega,t) = y(t) - \Gamma(\omega)$$

in which $\Gamma(\omega)$ is the total traveling cost (in the term of the industrial good) between dwelling site ω and the CBD. For simplicity of analysis, we specify $\Gamma(\omega)$ by

$$\Gamma(\omega) = A\omega^b, \quad A > 0, \quad 0 < b < 1, \quad 0 \le \omega < L. \tag{12.1.13}$$

Our assumption about the transportation system is very simplified. A more reasonable way to analyze effects of the transportation system upon the urban pattern is to introduce an endogenous transportation sector that takes part in economic competition. At this initial stage of the study, we are not concerned with these possibly too complicated issues about endogenous transportation systems.

The unique optimal solution is given by

$$c_a = \mu \rho y_c, \quad c_s = \frac{\gamma \rho y_c}{p_s}, \quad c_i = \frac{\xi \rho y_c}{p_i}, \quad c_h = \frac{\eta \rho y_c}{R}, \quad 0 \le \omega < L \tag{12.1.14}$$

in which

$$\rho \equiv \frac{1-s}{2\eta}.$$

Similarly, we define farmers' decision problem as follows

$$Max\ U(c_a, c_s, c_i, c_h)$$

subject to

$$c_a + p_s c_s + p_i c_i + R c_h = y_a, \quad L \le \omega < L^* \tag{12.1.15}$$

where

$$y_a(\omega, t) = y(t) - \Gamma_a(\omega)$$

in which $\Gamma_a(\omega)$ is the total traveling cost of the farmer between dwelling site ω and the CBD. As farmers' travel frequency to the CBD is dependent upon consumption level of services, we specify $\Gamma_a(\omega)$ by

$$\Gamma_a(\omega) = A_a c_s \left(\frac{L}{2} + \frac{L^*}{2}\right)^b, \quad A_a > 0, \ 0 < b < 1, \ L \le \omega < L^*. \tag{12.1.16}$$

We assume that the impact of distance upon farmers' travel cost to the CBD may be approximately measured by the average distance from the agricultural area to the CBD. As the farmers are charged by the same travel cost, this implies that the land rent is identical for $L \le \omega < L^*$.

The unique optimal solution is given by

$$c_a = \mu \rho y_c, \quad c_s = \frac{\gamma \rho y_c}{p_s}, \quad c_i = \frac{\xi \rho y_c}{p_i}, \quad c_h = \frac{\eta \rho y_c}{R}, \quad L \le \omega < L^* . \tag{12.1.17}$$

The population and consumption constraints

We denote $n(\omega, t)$ the residential density at dwelling site ω. According to the definitions, we have

$$n(\omega, t) = \frac{1}{c_s(\omega, t)}, \quad 0 \le \omega < L^* . \tag{12.1.18}$$

The population constraint of urban residents is given by

$$2\pi \int_0^L \omega n(\omega)\, d\omega = N_i + N_s . \tag{12.1.19}$$

The consumption constraints are given by

$$\int_0^L \omega n c_j \, d\omega + \int_L^{L^*} \omega n c_j \, d\omega = \frac{C_j}{2\pi}, \quad j = a, i, s \tag{12.1.20}$$

where C_j is the consumption level of product of sector j.

Capital accumulation

Denote s ($0 < s < 1$) the savings rate of the population out of the current gross income. For simplicity, we assume that the savings rate is constant. The national savings is given by: sY. The capital accumulation is thus formulated by

$$\frac{dK}{dt} = \frac{sY}{p_i} - \delta_k K \tag{12.1.21}$$

in which δ_k is the given depreciation rate of capital.

Since the industrial product consists of the investment and consumption, we have

$$C_i = F_i - \frac{sY}{p_i}, \quad C_i \geq 0. \tag{12.1.22}$$

Knowledge growth
Different processes, such as learning by doing, R&D activities, division of labor, have been taken into account as sources of knowledge accumulation in the literature of economic growth with endogenous knowledge (e.g., Romer, 1987, Grossman and Helpman, 1991, Zhang, 1991b, 1991c). In this chapter, we only take account of learning by doing in modeling knowledge accumulation. We propose knowledge growth

$$\frac{dZ}{dt} = \frac{\tau_a F_a}{Z^{\varepsilon_a}} + \frac{\tau_i F_i}{Z^{\varepsilon_i}} + \frac{\tau_s F_s}{Z^{\varepsilon_s}} - \delta_z Z \tag{12.1.23}$$

in which δ_z (≥ 0) is the depreciation rate of knowledge, τ_a, τ_i, τ_s, ε_a (≥ 0), ε_i (≥ 0), and ε_s (≥ 0) are parameters. Equation (12.1.23) implies that knowledge accumulation is positively related to production scales and there are scale effects in knowledge accumulation. Further explanation of the equation is referred to Zhang (1991d). For simplicity of analysis, we assume that $\varepsilon_a = 0$. We neglect possible contribution of the agricultural sector to knowledge accumulation.

Full employment of the labor force and capital
The assumption that the labor force and capital are fully employed yields

$$N_a + N_i + N_s = N, \quad K_a + K_i + K_s = K. \tag{12.1.24}$$

We have thus built the urban model describing dynamic spatial and sectorial distribution of the labor force, capital and land with endogenous capital and knowledge accumulation. There are 26 endogenous variables, K, K_a, K_i, K_s, N_a, N_i, N_s, L, Z, F_a, F_i, F_a, $c_a(\omega)$, C_a, $c_i(\omega)$, C_i, $c_s(\omega)$, C_s, $c_h(\omega)$, p_i, p_s, r, $R(\omega)$, w, U, and Y. The system consists of 26 independent equations. We now show that under appropriate conditions the system has solutions.

12.2 The Temporary Spatial Equilibrium

This section shows the existence of temporary spatial equilibrium. From

$$U(\omega_1) = U(\omega_2), \quad 0 \leq \omega_1 \leq \omega_2 < L$$

(12.1.11) and (12.1.14), we have

$$\frac{R(\omega_1)}{R(\omega_2)} = \left\{\frac{y_c(\omega_1)}{y_c(\omega_2)}\right\}^2, \quad 0 \le \omega_1 \le \omega_2 < L. \tag{12.2.1}$$

As

$$\frac{y_c(\omega_1)}{y_c(\omega_2)} \le 0, \quad 0 \le \omega_1 \le \omega_2 < L$$

from (12.2.1) we conclude that the urban land rent declines as the location is further away from the CBD.

As the utility level of an urban resident is equal to that of a farmer at $\omega = L$, from (12.1.11), (12.1.14) and (12.1.17), we have $y_c(L) = y_a(L)$. As

$$y_c(L) = y_a(L)$$

from (12.1.14) and (12.1.17) we can conclude that urban residents and farmers have identical consumption components at the boundary, i.e., c_a, c_i, c_s, and c_h are identical among the households at $\omega = L$. From $y_c(L) = y_a(L)$ we also get

$$c_s(L) = \frac{A}{A_a}\left(\frac{2L}{L + L^*}\right)^b, \quad \text{for a farmer.} \tag{12.2.2}$$

This implies that the consumption level of services by the farmers at the boundary is only dependent of the distance from farmers' dwelling sites to the CBD. The constant, A / A_a, is determined by transportation conditions.

From (12.A.1.5), (12.A.1.6) and (12.A.1.10) in Appendix 12.A.1, we can get the following equation determining the boundary, $L(t)$, between the residential area and the agricultural area

$$\Phi(L) \equiv \pi(N_i + N_s)\frac{y_c^2 + 2yb_2L^b - b_1AL^{2b}}{y_c + b_2L^b} + \frac{y_cN_a}{2} - Y = 0. \tag{12.2.3}$$

The problem is to find a positive L $(0 < L < L^*)$ such that $\Phi(L) = 0$ for a given positive Y. Obviously, $y_c(L)$ has to be positive, i.e.

$$y_c(L) = (1 - s)\frac{Y}{N} - AL^b > 0.$$

Define $L_0 \equiv (y / A)^b$. We have

$$y_c(L_0) = 0.$$

Utilizing $y_c(L) = (1 - s)Y / N - AL^b$ and (12.1.16), we have

$$\Phi(0) = (2\pi N_i + 2\pi N_s - N_a - 2N)(1 - s)\frac{Y}{2N}$$
$$= \{2(\pi - 1)\beta_i s_0 + 2(\pi - 1)\gamma\beta_s - \beta_a\mu\}(1 - s)\frac{s_2 Y}{2},$$
$$\Phi(L_0) = \left\{\frac{(1 - s)(N_i + N_s)b\pi}{(1 + b)N} - 1\right\}Y \tag{12.2.4}$$

where

$$s_0 \equiv \xi + \frac{s\eta}{1 - s}, \quad s_2 \equiv \frac{N}{\beta_a\mu + \beta_s\gamma + \beta_i s_0}.$$

It is reasonable to assume $\Phi(0) > 0$. If, for instance, $\beta_s\gamma > \mu\beta_a / 4$, we have $\Phi(0) > 0$. We further assume that $\Phi(L_0) < 0$, i.e.

$$\frac{(1 - s)(N_i + N_s)b\pi}{(1 + b)N} < 1.$$

This requirement is satisfied, for instance, when $s = 1/4$ and $b = 2/3$. As $\Phi(0) > 0$ and $\Phi(L_0) < 0$, for any positive Y, there is always a positive L $(0 < L < L_0)$ such that $\Phi(L) = 0$ and $y_c(L) > 0$. From $y_c(L) > 0$, we have

$$0 < L < \min(L_0, L^*).$$

Lemma 12.2.1.
Assume

$$\beta_i s_0 + \gamma\beta_s > \frac{\beta_a \mu}{2(\pi - 1)}, \quad \frac{(1 - s)(N_i + N_s)b\pi}{(1 + b)N} < 1.$$

For any given positive Y, there is always (at least) one urban equilibrium pattern. The boundary between the urban residential and the agricultural area is determined by (12.2.3) as a function of Y.

Another important question is whether $\Phi(L) = 0$ has a unique solution. This can be proved by examining the signs of $\Phi'(L)$ and $\Phi''(L)$ for $0 < L < \min(L_0, L^*)$. We omit the discussion as the expressions of $\Phi'(L)$ and $\Phi''(L)$ are too complicated to get explicit conclusion. But it is quite possible that the system has multiple temporary equilibrium urban patterns.

As shown in the appendix, we can determine capital and labor distribution among the three sectors as functions of K and N at any point of time as follows

$$K_a = \alpha_a \mu s_1 K, \quad K_s = \alpha_s \gamma s_1 K, \quad K_i = \alpha_i s_0 s_1 K, \tag{12.2.5}$$

$$N_a = \beta_a \mu s_2, \quad N_s = \beta_s \gamma s_2, \quad N_i = \beta_i s_0 s_2 \tag{12.2.6}$$

in which

$$s_1 \equiv \frac{1}{\alpha_a \mu + \alpha_i s_0 + \alpha_s \gamma}.$$

These equations simply imply that capital and labor distribution among the three sectors are linearly proportional to the total capital and the total labor, respectively. As shown in Appendix A.1, we can similarly determine the other variables as functions of K, Z and Y in the system.

12.3 The Dynamics of Capital and Knowledge

The previous section showed the existence of temporary equilibrium spatial patterns. This section will find out the dynamics of urban growth in terms of K and Z.

Using (12.1.1), (12.1.4), (12.1.7), (12.2.5), (12.2.6), and $Y / p_i = \eta F_i / s_0(1 - s)$, we can write the dynamics of capital and knowledge in the following form

$$\frac{dK}{dt} = \frac{s\eta F_i}{(1-s)s_0} - \delta_k K,$$

$$\frac{dZ}{dt} = \frac{\tau_i F_i}{Z^{\varepsilon_i}} + \frac{\tau_s F_s}{Z^{\varepsilon_s}} - \delta_z Z \tag{12.3.1}$$

in which F_i and F_s are functions of K and Z given by

$$F_s = A_s Z^{m_s} K^{\alpha_s}, \quad F_i = A_i Z^{m_i} K^{\alpha_i} \tag{12.3.2}$$

where

$$A_s = \gamma(\alpha_s s_1)^{\alpha_s} (\beta_s s_2)^{\beta_s}, \quad A_i = s_0(\alpha_i s_1)^{\alpha_i} (\beta_i s_2)^{\beta_i}.$$

Introduce

$$x_i \equiv \frac{m_i}{\beta_i} - \varepsilon_i - 1, \quad x_s \equiv \frac{\alpha_s m_i}{\beta_i} + m_s - \varepsilon_s - 1.$$

The following proposition provides the conditions for the existence of equilibria and for stability.

Proposition 12.3.1.
(1) If $x_i < 0$ and $x_s < 0$, the system has a unique stable equilibrium; (2) If $x_i > 0$ and $x_s > 0$, the system has a unique unstable equilibrium; and (3) If $x_i > 0$, $x_s < 0$ or $x_i < 0$, $x_s > 0$, the system has either no equilibrium or two equilibria. When the system has two equilibria, the one with higher values of K and Z is unstable and the other one is stable.

Possibility of multiple equilibria implies that the same economy may enter either poverty traps or enjoy prosperity, depending on, for instance, initial conditions. To interpret the proposition, it is sufficient for us to examine the stable case: $x_i < 0$ and $x_s < 0$, i.e.

$$\frac{m_i}{\beta_i} < \varepsilon_i + 1, \quad \frac{\alpha_s m_i}{\beta_i} + m_s < \varepsilon_s + 1.$$

Since m_i and m_s are respectively the efficiency of knowledge utilization of the industrial and service sector, an increase in efficiency of knowledge utilization tends to destabilize the system. On the other hand, as ε_i and ε_s are the measurements of scale effects in knowledge accumulation by the industrial and service sector, respectively, an increase in the decreasing scale effects tends to stabilize the system. We see that whether the system is stable or unstable depends upon the combined conditions of the economic structure and knowledge accumulation characteristics.

The remainder of this chapter carries out comparative statics analysis under presumed stability, i.e., $\Phi^* > 0$, where Φ^* is defined in (12.A.2.8).

12.4 The Population and the Spatial Structure

Taking the derivatives of (12.A.2.12) and (12.A.2.3) with respect to N, we obtain the effects of changes in the total labor force upon K and Z as follows

$$N\Phi^* \frac{dZ}{dN} = \frac{\tau_i F_i}{Z^{\varepsilon_i}} + \frac{\beta_s \tau_s F_s}{Z^{\varepsilon_s}} > 0, \quad \frac{dK}{dN} = \frac{m_i K}{\beta_i N} \frac{dZ}{dN} + \frac{K}{N} > 0. \tag{12.4.1}$$

As the labor force is increased, the equilibrium values of capital and knowledge tend to be expanded. It is easy to see that capital per capita is also increased, i.e., $d(K/N)/dN > 0$. One can directly check that N_a, N_i, N_s, K_a, K_i, K_s, F_i, F_s are increased as the labor force is increased. Here, it should be remarked that the above conclusion is dependent upon the assumption of $\Phi^* > 0$. If the system is unstable, i.e., $\Phi^* < 0$, then we may get the opposite conclusion about the impact of changes in N. Accordingly, the effects of changes in N upon the system are sensitive to whether the system is stable or unstable.

From (12.1.2) and (12.1.5), we have

$$p_s = \frac{\beta_a N_s F_a}{\beta_s N_a F_s}$$

where

$$F_a = A_a Z^{m_a} K^{\alpha} L_a^{\varsigma}, \quad A_a \equiv (\alpha_a \mu s_1)^{\alpha_a} (\beta_a \mu s_2)^{\beta_a}.$$

From (12.1.14) and (12.2.2), we have

$$p_s = \frac{\xi \rho A_a y_c(L)}{A} \left(\frac{L + L^*}{2L} \right)^b .$$

We also have

$$L_a + c_h(L) N_a = 2\pi \left(L^{*2} - L^2 \right), \quad c_h(L) = \frac{\eta \rho y_c}{R} = \frac{\eta \rho y_c L_a}{\varsigma F_a} .$$

With these relations and (12.2.3), we have

$$\left(y_c^2 + 2 y b_2 L^b - A b_1 L^{2b} \right) \pi_0 + \left(y_c - n_0 y \right)\left(y_c + b_2 L^b \right) = 0,$$

$$m_a \left(1 + \frac{\varsigma_0 N_a y_c}{F_a} \right) = 2\pi \left(L^{*2} - L^2 \right), \quad y_c \left(1 + \frac{L^*}{L} \right)^b = \frac{b_0 F_a}{F_s} \qquad (12.4.2)$$

in which

$$y_c = y - A L^b, \quad \pi_0 \equiv 2\pi \frac{N_i + N_s}{N_a}, \quad n_0 \equiv \frac{2N}{(1 - s) N_a},$$

$$b_0 \equiv \frac{2^b A \beta_a N_s}{\beta_s \xi \rho A_a N_a}, \quad \varsigma_0 \equiv \frac{\eta \rho}{\varsigma}.$$

There are 5 variables, K, Z, y, L_a and L in (12.4.1). As (12.4.1) gives dK/dN and dZ/dN, it is possible to directly determine dy/dN, dL_a/dN and dL/dN from (12.4.2).

Taking derivatives of (12.4.2) with respect to N yields

$$\left\{ (2\pi_0 + 1 - n_0) y + \frac{4\pi_0 b_2}{b} + A L^b \right\} \frac{dy}{dN} = \left\{ A b y_c + A b b_2 L^b + \right.$$

$$\left. 2 b y_c + 2 y b_2 (2 b \pi_0 + 3\pi_0 - n_0) - 2\pi_0 A b L^b (2A - b_1) \right\} L^{b-1} \frac{dL}{dN},$$

$$\frac{dy}{dN} = \left(AbL^{b-1} - \frac{4\pi LF_a}{\varsigma_0 L_a N_a} \right) \frac{dL}{dN} - \left(\frac{F_a}{\varsigma_0 L_a N_a} + \frac{1-\varsigma}{L_a} y_c \right) \frac{dL_a}{dN}$$

$$+ \left(F^* - \frac{1 - m_a}{N} \right) y_c,$$

$$\frac{dy}{dN} = a_0 y_c - \frac{\varsigma y_c}{L_a} \frac{dL_a}{dN} - \left(AbL^{b-1} + \frac{L^* y_c / L}{L^* + L} \right) \frac{dL}{dN} \tag{12.4.3}$$

in which

$$F^* \equiv \left(\frac{\alpha_a m_i}{\beta_i Z} + \frac{m_a}{Z} \right) \frac{dZ}{dN} + \frac{\alpha_a}{N},$$

$$a_0 \equiv \frac{1 + \alpha_a - 2\alpha_s}{N} + \left\{ \frac{m_a - m_s + m_i (\alpha_a - \alpha_s) / \beta_i}{Z} \right\} \frac{dZ}{dN}.$$

For simplicity of discussion, we assume $y_c \geq 2AL^b$, i.e., $y \geq 3AL^b$. This implies that the consumer dwelling near the boundary between the urban and agricultural area spends less than one-third of his income on traveling to the CBD. As this requirement is not strict, we accept it in the remainder of this section. We assume the existence of meaningful parameter values such that the requirement is satisfied. We also assume $2b\pi_0 - n_0 \geq 0$, i.e.

$$\frac{N_i + N_s}{N} > \frac{1}{4\pi b(1 - s)}.$$

This simply means that the share of the sum of the labor force employed in the industrial and service sectors in the total labor force is not very small (e.g., $(N_i + N_s)/N > 1/6$ in the case of $s = 1/4$ and $b = 2/3$). Under these requirements, we see the term before dL / dN in the first equation in (12.4.3) is positive. That is

$$Aby_c + Abb_2 L^b + 2yb_2 (2b\pi_0 + 3\pi_0 - n_0) - 2\pi_0 AbL^b (2A - b_1)$$
$$+ 2by_c > [3Ab + (4 + b)b_2 + 2\pi_0 \{2b_2 (1 - b) + bb_1\}] AL^b > 0.$$

Accordingly, from the first equation in (12.4.3) we have

$$\frac{dL}{dN} = a^* \frac{dy}{dN} \tag{12.4.4}$$

where a^* is a positive constant (evaluated at the equilibrium).

Substituting (12.4.4) into the second and third equations in (12.4.3), we get the impact upon the income per capita as follows

$$\left[1 - a^* AbL^{b-1} + \left(2b^* + \frac{1}{\varsigma} - 1\right)\left\{1 + a^* AbL^{b-1} + \frac{a^* L^* y_c / L}{L^* + L}\right\}\right.$$
$$\left. + \frac{4\pi a^* LF_a}{\varsigma_0 L_a N_a}\right]\frac{dy}{dN} = \{2b^*(1 + m_a - 2\alpha_s) - 1 + m_a + \alpha_a\}\frac{y_c}{N} +$$
$$\left[\left\{m_a - m_s + \frac{(m_a - \alpha_s)m_i}{\beta_i}\right\}2b^* + \frac{m_a m_i}{\beta_i} + m_a\right]\frac{y_c}{Z}\frac{dZ}{dN}$$
$$+ \left(\frac{1}{\varsigma} - 1\right)a_0 y_c \tag{12.4.5}$$

in which

$$b^* \equiv \frac{F_a}{(1 - s)y_c N_a}\left(= \frac{C_a}{(1 - s)y_c N_a} = \frac{\mu y N}{(1 - s)\eta y_c N_a}\right).$$

The term before dy/dN is positive when $2b^* \geq 1$. Examining (12.4.5), we have the following proposition.

Proposition 12.4.1.
Let $2b^* \geq 1$, $1 + m_a \geq 2\alpha_s$, $-2b^*(1 + \alpha_a - 2\alpha_s) \leq m_a + \alpha_a - 1 \leq 0$, and $(1/Z)dZ/dN$ be small. Then, as the labor force is increased, the income per capita, y, will be increased and the boundary will be moved outward as the population is growing.

From (12.4.5), we see that it is reasonable for dy/dN and dL/dN to be negative. For instance, when b^* and $(1/Z)dZ/dN$ are large, we may get the conclusions opposite to these in Proposition 12.4.1.

From the second and third equations, we get

$$\left(\frac{F_a}{\varsigma_0 L_a N_a} + \frac{y_c}{L_a}\right)\frac{dL_a}{dN} = \left\{\left(\frac{m_i}{\beta_i} + 1\right)\frac{m_a}{Z}\frac{dZ}{dN} - \frac{1 - m_a - \alpha_a}{N}\right\}y_c$$

$$+ \left(2AbL^{b-1} - \frac{8\pi b^* \varsigma L y_c}{L_a} + \frac{L^* y_c / L}{L^* + L}\right)\frac{dL}{dN}. \qquad (12.4.6)$$

If the assumptions in Proposition 12.4.1 are held, then we have $dL_a / dN < 0$. Hence, as the labor force is increased, the income per capita, the boundary are increased, but the land utilized in the agricultural production declines.

From the second equation in (12.4.3) and (12.4.6), we have

$$\frac{dy_c}{dN} = \frac{dy}{dN} - AbL^{b-1}\frac{dL}{dN} = \left\{\left(1 - 2\varsigma^*\right)AbL^{b-1} - \frac{\left(1 - \varsigma^*\right)8\pi b^* \varsigma L y_c}{L_a}\right.$$

$$\left. - \frac{L^* \varsigma^* y_c / L}{L^* + L}\right\}\frac{dL}{dN} + \left\{\left(\frac{m_i}{\beta_i} + 1\right)\frac{m_a}{Z}\frac{dZ}{dN} - \frac{1 - m_a - \alpha_a}{N}\right\}y_c.$$

$$(12.4.7)$$

where

$$\varsigma^* \equiv \frac{F_a / \varsigma_0 N_a + (1 - \varsigma)y_c}{F_a / \varsigma_0 N_a + y_c} = 1 - \frac{\varsigma}{1 + 2\varsigma b^*} < 1.$$

We see that $y_c(L)$ may be either positive or negative, depending upon the parameter values.

Utilizing $y_c(\omega) = y - \Gamma(\omega)$ and (12.2.2), we have

$$\frac{dy_c(\omega)}{dN} = \frac{dy}{dN}, \quad \frac{1}{c_s(L)}\frac{dc_s(L)}{dN} = \frac{L^*}{L}\frac{dL}{dN} > 0. \qquad (12.4.8)$$

We now show how, utilizing the above analytical results, we can analyze effects of changes in N upon the other variables. Utilizing the following relations

$$Y = \frac{(1 - s)y}{N}, \quad R = \frac{\varsigma F_a}{L_a}, \quad p_s = \frac{\xi \rho y_c(L)}{c_c(L)}, \quad r = \frac{\alpha_s p_s F_s}{K_s},$$

$$w = \frac{\beta_s p_s F_s}{N_s}, \quad p_i = \frac{wN_i}{\beta_i F_i} \tag{12.4.9}$$

and (12.1.1), (12.1.4) (12.1.7), (12.1.11), (12.1.14) and (12.1.17), we can easily get the impact of changes in N upon the 26 endogenous variables, K, K_a, K_i, K_s, N_a, N_i, N_s, L, Z, F_a, F_i, F_s, $c_a(\omega)$, C_a, $c_i(\omega)$, C_i, $c_s(\omega)$, C_s, $c_h(\omega)$, p_i, p_s, r, $R(\omega)$, w, U, and Y in the system.

Similarly, we may examine the impact of changes in the other parameters.

12.5 On Complexity of Economic Geography

This chapter suggested an urban model to describe the dynamic mechanisms of regional growth with endogenous spatial pattern formation. We were concerned with an idealized economic system consisting of three kinds of economic activities with homogenous labor force. We specified the location of industrial and service sectors at the CBD, the residential location of urban workers in the residential area and the location of farmers in the agricultural area. The dynamic system consists of two-dimensional differential equations of capital and knowledge accumulation with economic spatial structure determined as functions of capital and knowledge at any point of time. We provided the conditions for existence of temporary and long-run equilibria and the conditions for stability of the system.

For simplicity of analysis, we neglected the land rent from housing market. This assumption may be relaxed either by introducing absentee landlords, or by including the rent explicitly in the gross income equation. We did not distinguish any essential difference in consumption between urban and rural households. For simplicity of analysis, we also made other assumptions, such as $\gamma + \xi + \mu = \eta$, the savings rate being constant and τ_a, which are obviously rather limited cases. We also neglect transportation costs of agricultural and industrial product to consumers. The introduction of such costs will cause more analytical difficulties. These assumptions can be, in principle, relaxed, though further complexity of analysis will be resulted. It is also possible to extend the model to an economic system consisting of multiple regions.

This chapter is influenced by Thünen's economic system, Alonso's urban system and the contemporary growth theory (with endogenous knowledge). As our system is a synthesis of the three different approaches, it is possible to extend our model along the lines that these three approaches have been developed. The literature on urban economics points out many obvious directions to further examine the complexity of the residential area. Thünen's system implies the limitation and invalidity of our simplified agricultural area. There are many studies of economic growth theory which help us to take account of other aspects of economic reality. International and regional economics may shed light on how we can extend the work to multiple regions.

Appendix

A.12.1 Proving the Results in Section 12.2

We now derive (12.2.5) and (12.2.6). As the agricultural area consists of dwelling sites and land for farming, i.e., $L_a + c_h(L)N_a = 2\pi\left(L^{*2} - L^2\right)$, we have

$$N_a = 2\pi\frac{L^{*2} - L^2}{c_h} - \frac{\varsigma C_a}{c_h R} \tag{12.A.1.1}$$

in which we use $L_a = \varsigma F_a / R = \varsigma C_a / R$. It should be remarked that c_h and R in (12.A.1.1) take on their values at the boundary, i.e., $c_h = c_h(L)$ and $R = R(L)$. Similarly, in the remainder of this section all spatial variables, if without special mention, are evaluated at the boundary $\omega = L$.

Utilizing (12.1.14) and (12.2.1), we have

$$n(\omega) = \frac{1}{c_h(\omega)} = \frac{R(\omega)}{\eta\rho y_c(\omega)} = \frac{y_c(L)R(L)}{\eta\rho y_c^2(\omega)}, \quad 0 < \omega < L. \tag{12.A.1.2}$$

From (12.1.20), we get

$$C_a = c_a N_a + 2\pi\int_0^L n(\omega)c_a(\omega)\omega\, d\omega = \mu Y^* \tag{12.A.1.3}$$

where

$$n(\omega)c_a(\omega) = \frac{c_a(\omega)}{c_h(\omega)} = \frac{\mu}{\eta}R(\omega) = \mu y_c^2(\xi)Y_c, \quad 0 \le \omega < L, \tag{12.A.1.4}$$

$$Y^* \equiv \left(y_c^2 + 2yb_2L^b - Ab_1L^{2b}\right)\pi L^2 Y_c + y_c \rho N_a \tag{12.A.1.5}$$

in which

$$Y_c \equiv \frac{R}{\eta y_c^2(L)}, \quad b_1 \equiv \frac{bA}{1+b}, \quad b_2 \equiv \frac{bA}{2+b}.$$

Substituting (12.A.1.3) into (12.A.1.1) yields

$$N_a = \frac{\left(L^{*2} - L^2\right)2\pi R - \varsigma\mu Y^*}{\eta\rho y_c}. \tag{12.A.1.6}$$

From (12.1.19) and (12.A.1.2), we have

$$N_i + N_s = \frac{y_c + b_2L^b}{2\rho} Y_c L^2. \tag{12.A.1.7}$$

By

$$F_i = \frac{sY}{p_i} + C_i, \quad F_a = C_a, \quad F_s = C_s$$

we have

$$(1-s)Y = C_a + p_iC_i + p_sC_s. \tag{12.A.1.8}$$

From (12.1.14) and (12.1.17), we have the following relations

$$p_iC_i = p_ic_iN_a + 2\pi p_i \int_0^L n(\omega)c_i(\omega)\omega\, d\omega = \xi Y^*,$$

$$p_sC_s = p_sc_sN_a + 2\pi p_s \int_0^L n(\omega)c_s(\omega)\omega\, d\omega = \gamma Y^*,$$

$$2\pi \int_0^L R(\omega)c_h(\omega)\omega\, d\omega = \left(y_a + b_2L^b\right)\pi\eta\rho L^2. \tag{12.A.1.9}$$

Substituting (12.A.1.3) and (12.A.1.8) into (12.A.1.8) yields

$$\eta Y^* = (1 - s)Y . \tag{12.A.1.10}$$

We now show that N_a, N_i and N_s are proportional to N. From (12.1.2), (12.1.5) and (12.1.8), we have

$$\frac{K_i}{K_a} = \frac{\alpha_i \beta_a}{\alpha_a \beta_i} \frac{N_i}{N_a}, \quad \frac{K_i}{K_s} = \frac{\alpha_i \beta_s}{\alpha_s \beta_i} \frac{N_i}{N_s} . \tag{12.A.1.11}$$

From

$$r = \frac{\alpha_i p_i F_i}{K_i} = \frac{\alpha_s p_s C_s}{K_s} = \frac{\alpha_a C_a}{K_a}, \quad C_a = \mu Y^*, \quad p_i C_i = \xi Y^*,$$

$$p_s C_s = \gamma Y^*, \quad F_i = C_i + \frac{sY}{p_i}, \quad (1 - s)Y = \eta Y^*$$

we have

$$\frac{K_a}{K_s} = \frac{\alpha_a \mu}{\alpha_s \gamma}, \quad \frac{K_a}{K_i} = \frac{(1 - s)\alpha_a \mu}{\{(1 - s)\xi + s\eta\}\alpha_i} . \tag{12.A.1.12}$$

Substituting (12.A.1.12) into $K_a + K_i + K_s = K$ yields (12.2.5). From (12.A.1.11), (12.A.1.12), and $N_a + N_i + N_s = N$, we get (12.2.6).

A.12.2 Proving Proposition 12.3.1

The long-run equilibrium is given by

$$\frac{s\eta F_i}{(1 - s)s_0} = \delta_k K , \quad \frac{\tau_i F_i}{Z^{\varepsilon_i}} + \frac{\tau_s F_s}{Z^{\varepsilon_s}} = \delta_z Z . \tag{12.A.2.1}$$

From the first equation, we have

$$K = A Z^{m_i / \beta_i} \tag{12.A.2.2}$$

in which

$$A = \left\{ \frac{s\eta A_i}{\delta_k s_0 (1 - s)} \right\}^{1/\beta_i} > 0 .$$

Substituting (12.A.2.2) into the second equation in (12.A.2.1) yields

$$H(Z) = \Phi_i(Z) + \Phi_i(Z) - \delta_z = 0 \qquad (12.A.2.3)$$

where

$$\Phi_i(Z) \equiv \tau_i A_i A^{\alpha_i} Z^{x_i}, \quad \Phi_s(Z) \equiv \tau_s A_s A^{\alpha_s} Z^{x_s} .$$

We now try to find conditions such that $H(Z) = 0$ has positive solutions. We exclude the case of $x_i = x_s = 0$. If $x_i \geq 0$ and $x_s \geq 0$ (i.e., $H(0) < 0$, $H(\infty) > 0$ and $H' > 0$ for $Z > 0$) or $x_i \leq 0$, $x_s \geq 0$ (i.e., $H(0) > 0$, $H(\infty) < 0$ and $H' < 0$ for $Z > 0$), then the system has a unique positive equilibrium. If i) $x_i > 0$, $x_s < 0$, ii) $x_i < 0$, $x_s > 0$, then the system has either two equilibria or no one. We just prove case i). Case ii) can be similarly checked. Since $x_i > 0$ and $x_s < 0$, $H(0) > 0$ and $H(\infty) > 0$. This implies that $H(Z) = 0$ has either no solution or multiple solutions. Since

$$ZH'(Z) = x_i \Phi_i + x_s \Phi_s \qquad (12.A.2.4)$$

we see that $H'(Z)$ may be either positive or negative, depending upon the parameter values. If $H(Z) = 0$ has more than two solutions, there are at least two values of Z such that $H'(Z) = 0$ (i.e., $ZH'(Z) = 0$). Since $dZH'/dZ > 0$ strictly holds for $Z > 0$, it is impossible for $H'(Z) = 0$ to have more than one solution. Accordingly, $H(Z) = 0$ has either no solution or two solutions. A necessary and sufficient condition for the existence of two equilibria is that there exists a value of Z^* such that $H(Z^*) < 0$ and $H'(Z^*) = 0$. It is easy to check that when $H(Z) = 0$ has two solutions, Z_1 and Z_2 ($Z_1 > Z_2 > 0$), $H'(Z_1) > 0$ and $H'(Z_2) < 0$. This is important for proving the stability of the equilibria.

We now provide the stability conditions for equilibria. The Jacobian at equilibrium is given by

$$J = \begin{pmatrix} -(1-\alpha_i)\delta_k & \dfrac{\delta_k m_i K}{Z} \\ \dfrac{a_1}{K} & -\dfrac{a_2}{Z} \end{pmatrix} \tag{12.A.2.5}$$

in which

$$a_1 \equiv \frac{\alpha_i \tau_i F_i}{Z^{\varepsilon_i}} + \frac{\alpha_s \tau_s F_s}{Z^{\varepsilon_s}} > 0,$$

$$a_2 = (1 + \varepsilon_i - m_i)\frac{\tau_i F_i}{Z^{\varepsilon_i}} + (1 + \varepsilon_s - m_s)\frac{\tau_s F_s}{Z^{\varepsilon_s}}. \tag{12.A.2.6}$$

The two eigenvalues, ϕ_1 and ϕ_2, are determined by

$$\phi^2 + \left(\delta_k \beta_i + \frac{a_2}{Z}\right)\phi + \frac{a_2 \delta_k \beta_i}{Z} - \frac{a_1 \delta_k m_i}{Z} = 0. \tag{12.A.2.7}$$

Accordingly, if $\beta_i a_2 < m_i a_1$, the system is unstable; if $\beta_i a_2 > m_i a_1$, the system is stable. The condition, $\beta_i a_2 > m_i a_1$, is satisfied if

$$\Phi^*(Z) \equiv -\frac{x_i \Phi_i + x_s \Phi_s}{Z} > 0. \tag{12.A.2.8}$$

Summarizing the above discussion, we proved the proposition.

13 Further Issues on Interregional Economics

The whole, however, is merely the essential nature reaching its completeness through the process of its own development.

Hegel: The Phenomenology of Mind (1807)

In my previous books (Zhang, 1999, 2000, 2001), I endeavored to examine various aspects of national, urban, and international economies with a few concepts within a single framework. This book applied the basic concepts in these books to explore complex interdependence between economic growth, economic structures, and sexual division of labor within an interregional context. It explored the complexity of interregional trade within the same framework that I employed to examine national, urban, and international economies. Trying to say many things with as a few concepts and techniques as possible should be the manifestation of the bona fide mind of the economist. Heidegger (1993:271) once recounted: "'Are you still standing there,' condescendingly asked the much-traveled Sophist of Socrates, 'and still saying the same thing about the same thing?' 'Yes,' answered Socrates, "that I am."

There are many directions of refining and extending the models proposed in this book. Our focus on perfect competition is different from the contemporary mainstream of formal regional theory that is mainly concerned with monopolistic competition or fixed prices frameworks. It is necessary to examine the two sides of the same reality within an organic framework – this requires innovative ideas and new frameworks not to mention a much more refined analytical tools. As far as efficiency of modeling is concerned, one may advance knowledge rapidly and unflickeringly by focusing either on the *yin* or the *yang*.

Our book is focused on dynamics of economic structures, capital, and knowledge, and human capital. It is not intended to be a comprehensive treatment of all the important issues related to interregional trades. We did not discuss issues such as monetary variables and the role of expectations. Moreover, this book does not represent any model of 'an open economy'. Since this book is based on a comprehensive system of national economies, it is conceptually easy to make further extensions. We now mention a few directions of possible extensions of our work.

From the literature of classical economics we know at least four input factors which may exhibit increasing or decreasing returns to scale effects in economic dynamics:

infrastructures (of transportation and communication systems), institutions, knowledge and population, (e.g., Malthus, 1933, Marshall, 1890, Haavelmo, 1954, Niehans, 1963, Zhang, 1991d, 1996, 1999). In a long-term analysis, it is necessary to examine decision-making processes involved in construction and maintenance of infrastructures. Channels, roads, railways and airline systems, which may be effectively treated as parameters in short-term analyses, determine the mobility and the costs associated with movements of people and goods. This book introduced endogenous knowledge and capital. It is not difficult to introduce endogenous population and networks into our analytical framework.

We did not deal with economic evolution with knowledge in a comprehensive way because of analytical difficulties for obtaining explicit conclusions. Following the modeling framework developed by Zhang (1999), it is conceptually easy to introduce other aspects such as human capital structures and professional amenities into our framework. It is not difficult to relax the assumption of a single kind of capital (Burmester and Dobell, 1970, Takayama, 1985). The introduction of multiple capital goods will cause analytical difficulties. It should be noted that the traditional neoclassical growth theory did not succeed in dealing with growth issues with multiple capital goods in the sense that the consumer behavior was not properly modeled. Although we developed multi-group trade models, our classification of labor force was simplified. It should be noted that different kinds of labor force may enter production functions in different ways. We neglected issues related to natural resources and environment. The issues related to trade, resources and capital accumulation should be further examined.

We used simple production functions in this book. In economics, it is common to use more general forms of production functions. Moreover, it is reasonable to introduce other factor inputs such as natural resources and infrastructures into production functions. We assumed perfect competition in domestic markets. But imperfect competition is important to describe the economic reality. If markets are not perfectly competitive, prices of goods and services may not be rapidly changed as we assumed in the dynamic models. Utility functions may be taken on various forms (Benson, 1980, 1984, Berliant and Tee Raa, 1987, Cheshire and Evans, 1991). Except common issues related to forms of utility functions (Lancaster, 1966, 1971, Boyer, 1978), we may also introduce preference change. Furthermore, it should be remarked that Dixit and Stiglitz (1977) have emphasized the implications of the endogenization of the number of goods for productivity progress and R&D. In general this idea can be taken into account by assuming that knowledge affects the parameters in the utility and production functions.

We assumed that production factors such as labor force, capital and land are always fully employed. These assumptions should be relaxed in order to analyze modern economies. In particular, one of the central topics in interregional theory is related to the impact of interregional trade on unemployment. This issue can be handled according to the model with unemployment suggested in Zhang (1999).

Another important issue is to examine impact of international trade on interregional economics. We may consider a world consisting of m (for instance, $m = 3$) countries and country j is made up by n_j $(j = 1, 2, 3)$ regions. International economic interactions can be illustrated as in Fig. 12.1. With the analytical frameworks proposed in this book and Zhang (2001), the reader can descry that it is analytically not difficult to examine the economic geography illustrated in the figure. In fact, there are many possible networking structures of economic geography and each structure requires concrete examination. With help of contemporary computer, one may scrutinize dynamic behavior of differently detailed economic structures.

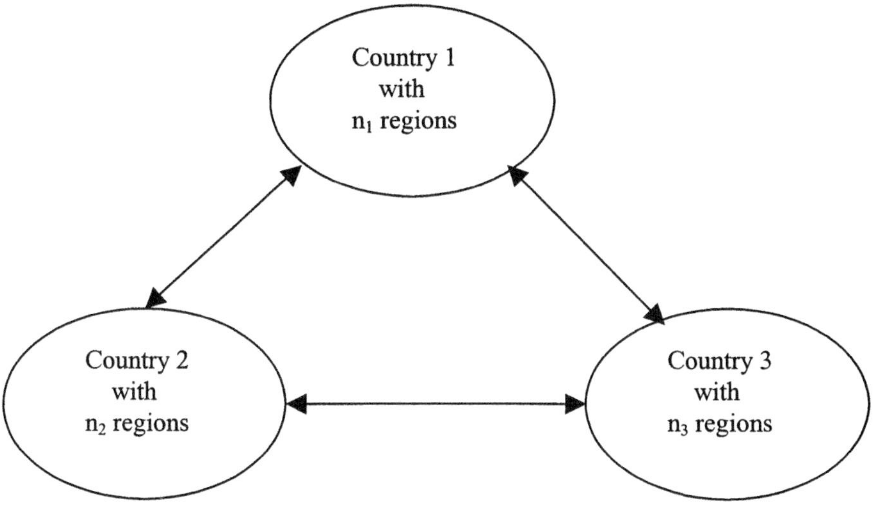

Fig. 12.1. International and Interregional Economies

As far as my knowledge is concerned, this is the first book on interregional economics in a theoretically comprehensive sense (even in the viewpoint of the equilibrium economics and the neoclassical growth theory). The number of questions raised in constructing such a comprehensive and complicated framework must be more than the number of questions fully answered. I wish that the book might provide a starting point for young researchers to build much more refined and comprehensive regional economic theories. Regional economics needs energetic and creative researchers. The scarcity of talents tells: rarely does beauty and practical usefulness come by in the same person. The skin depth of scientific beauty can be truly appreciated only by the deep mind. The telos of the economist is to produce economic theories that satisfy the heart as well as the mind, not mention to solve problems of economic life.

Bibliography

Abdel-Rahman, H.M. (1988): Product Differentiation, Monopolistic Competition and City Size. Regional Science and Urban Economics 18, 69-86

Aghion, P., Howitt, P. (1992): A Model of Growth through Creative Destruction. Econometrica 60, 323-351

Aghion, P., Howitt, P. (1998): Endogenous Growth Theory. Mass., Cambridge: The MIT Press

Alonso, W. (1964): Location and Land Use. MA., Cambridge: Harvard University Press

Anas, A. (1992): On the Birth and Growth of Cities - Laissez-Faire and Planned Compared. Regional Science and Urban Economics 22, 243-258

Andersson, Å.E. (1986): The Four Logistical Revolutions. Papers of Regional Science Association 59, 1-12

Andersson, Å.E., Anderstig, C., Hårsman, B. (1990): Knowledge and Communications Infrastructure and Regional Economic Change. Regional Science and Urban Economics 20, 359-376

Arnott, R. (1996a, Ed.): Regional and Urban Economics, Part 1. Amsterdam: Harwood Academic Publishers

Arnott, R. (1996b, Ed.): Regional and Urban Economics, Part 2. Amsterdam: Harwood Academic Publishers

Arrow, K.J. (1962): The Economic Implications of Learning by Doing. Review of Economic Studies 29: 155-173

Barro, R.J., Sala-i-Martin, X. (1995): Economic Growth. New York: McGraw-Hill, Inc

Batten, D.F. (1982): The Interregional Linkages Between National and Regional Input-Output Analysis. International Regional Science Review 7, 53-68

Becker, G.S. (1965): A Theory of the Allocation of Time. Economic Journal 75, 493-517

Becker, G.S. (1976): The Economic Approach to Human Behavior. Chicago: The University of Chicago Press

Becker, G.S. (1981): A Treatise on the Family. Mass., Cambridge: Harvard University Press

Becker, G.S. (1985): Human Capital, Effort, and the Sexy Division of Labor. Journal of Labor Economics 3, S34-58

Beckmann, M.J. (1957): On the Distribution of Rent and Residential Density in Cities, Paper presented at the inter-Departmental Seminar on Mathematical Applications in the Social Sciences, Yale University

Beckmann, M.J. (1968): Location Theory. New York: Harper & Brothers.

Beckmann, M.J. (1969): On the Distribution of Urban Rent and Residential Density. Journal of Economic Theory 1, 60-68

Beckmann, M.J. (1976): Spatial Equilibrium in a Dispersed City, in Mathematical Land Use Theory, edited by Y.Y. Papageorgiou. Lexington: Lexington Books

Beckmann, M.J. (1990): Taste and Welfare in Travellers' Choices. Studies in Regional Science 20, No-2, 25-29

Beckmann, M., Puu, T. (1985): Spatial Economics: Density, Potential, and Flow. Amsterdam: North-Holland

Bell, C. (1991): Regional Heterogeneity, Migration, and Shadow Prices. Journal of Public Economics 46, 1-27

Benson, B. (1980): Löschian Competition Under Alternative Demand Conditions. American Economic Review 70, 1098-105

Benson, B. (1984): Spatial Competition with Free Entry, Chamberlinian Tangencies, and Social Efficiency. Journal of Urban Economics 15, 270-86

Berliant, M., Ten Raa, T. (1987): On the Continuum Approach of Spatial and Some Local Public Goods or Product Differentiation Models. Working Paper 72, Rochester Center for Economic Research, University of Rochester

Blaug, M. (1985): Economic Theory in Retrospect, fourth edition. Cambridge: Cambridge University Press

Blomquist, G.C., Berger, M.C., Hoehn, J.C. (1988): New Estimates of Quality of Life in Urban Areas. American Economic Review 78, 89-107

Borts, G.H., Stein, J.L. (1964): Economic Growth in a Free Market. New York: Columbia University Press

Boyer, M. (1978): A Habit Forming Optimal Growth Model. International Economic Review 19, 585-609

Buiter, W.H. (1981): Time Preference and International Lending and Borrowing in an Overlapping Generations Model. Journal of Political Economy 89, 769-797

Burmeister, E., Dobell, A.R. (1970): Mathematical Theories of Economic Growth. London: Collier Macmillan Publishers

Cain, G.G. (1986): The Economic Analysis of Labor Market Discrimination: A Survey. In Ashenfelter, O.C., Layard, R. (Eds.): Handbook of Labor Economics. Amsterdam: North-Holland

Calem, P.S., Carlino, G.A. (1991): Urban Agglomeration Economies in the Presence of Technical Change. Journal of Urban Economics 29, 82-95

Cheshire, P.C., Evans, A.W. (1991, Eds.): Urban and Regional Economics. Hants: Edward Elgar Publishing Company

Chiappori, P.A. (1988): Rational Household Labor Supply. Econometrica 51, 63-89

Chiappori, P.A. (1992): Collective Labor Supply and Welfare. Journal of Political Economy 100, 438-467

David, P.A., Rosenbloom, J.L. (1990): Marshallian Factor Market Externalities and the Dynamics of Industrial Localization. Journal of Urban Economics 28, 349-370

Diamond, D.B., Tolley, G.S. (1981, Eds.): The Economics of Urban Amenities. New York: Academic Press

Dixit, A.K., Stiglitz, J.E. (1977): Monopolistic Competition and Optimum Product Diversity. American Economic Review 67, 297-308

Dixon, R.J., Thirlwall, A.P. (1975): A Model of Regional Growth Rate Differences on Kaldorian Lines. Oxford Economic Papers 2, 201-214

Dollar, D. (1986): Technological Innovation, Capital Mobility, and the Product Cycle in the North-South Trade. American Economic Review 76, 177-190

Dosi, G., Pavitt, K., Soete, L. (1990): The Economics of Technical Change and International Trade. New York: Harvester Whertsheaf

Drandakis, E., Phelps, E.S. (1966): A Model of Induced Invention, Growth and Distribution. Economic Journal LXXVI, 823-40

Englmann, F.C., Walz, U. (1995): Industrial Centers and Regional Growth in the Presence of Local Inputs. Journal of Regional Science 35, 3-27

Ethier, W.J. (1982): National and International Returns to Scale in the Modern Theory of International Trade. American Economic Review 72, 389-405

Fei, J.C.H., Ranis, G., Kuo, S.W.Y. (1978): Growth and the Family Distribution of Income by Factor Components. Quarterly Journal of Economics XCII, 17-53

Findlay, R. (1984): Growth and Development in Trade Models. In Jones, R.W., Kenen, R.B. (Eds.): Handbook of International Economics. Amsterdam: North-Holland

Folbre, N. (1986): Cleaning House - New Perspectives on Households and Economic Development. Journal of Economic Development 22, 5-40

Frenkel, J., Razin, A. (1987): Fiscal Policy and the World Economy. MA., Cambridge: MIT Press

Fujita, M. (1989): Urban Economic Theory - Land Use and City Size. Cambridge: Cambridge University Press

Fujita, M., Krugman, P., Venables, A.J. (1999): The Spatial Economy – Cities, Regions, and International Trade. Mass., Cambridge: The MIT Press

Greenhut, M.L., Norman, G., Hung, C.S. (1987): The Economics of Imperfect Competition - A Spatial Approach. Cambridge: Cambridge University Press

Grossman, G.M., Helpman, E. (1991): Innovation and Growth in the Global Economy. Cambridge, Mass.: The MIT Press

Haavelmo, T. (1954): A Study in the Theory of Economic Evolution. Amsterdam: North-Holland

Heckman, J.J., Macurdy, T.E. (1980): A Life Cycle Model of Labor Supply. Review of Economic Studies 47, 47-74

Heidegger, M. (1993): Basic Writings of Martin Heidegger. Edited by D.F. Krell. San Francisco: Harper

Helsley, R.W., Starnge, W.C. (1991): Agglomeration Economies and Urban Capital Markets. Journal of Urban Economics 29, 96-112

Helsley, R.W., Sullivan, A.M. (1991): Urban Subcenter Formation. Regional Science and Urban Economics 21, 255-76

Henderson, J.V. (1985): Economic Theories and Cities. 2nd edition, Academic Press, New York

Ikeda, S., Ono, Y. (1992): Macroeconomic Dynamics in a Multi-Country Economy - A Dynamic Optimization Approach. International Economic Review 33, 629-644

Isard, W. (1953): Some Empirical Results and Problems ofRegional Input-Output Analysis, in W. Leontief, et al., Studies in the Structure of the American Economy. New York: Oxford University Press

Isard, W. (1956): Location and Space Economy. MA., Cambridge: MIT Press
Isard, W. (1960): Methods of Regional Analysis. Mass., Cambridge: MIT Press
Johansson, B., Karlsson, C. (1990, Eds.): Innovation, Industrial Knowledge and Trade - A Nordic Perspective. Umeå: CERUM at University of Umeå
Kaldor, N. (1966): Marginal Productivity and the Macro-Economic Theories of Distribution. Review of Economic Studies 33, No. 4
Kanemoto, Y. (1980): Theories of Urban Externalities. Amsterdam: North-Holland.
Kennedy, C. (1964): Induced Bias in Innovation and the Theory of Distribution. Economic Journal LXXIV, September
Krugman, P. R. (1979): A Model of Innovation, Technology Transfer and the World Distribution of Income. Journal of Political Economy 87
Krugman, P.R. (1989): Increasing Returns, Monopolistic Competition, and International Trade. Journal of International Economics 9, 469-479
Krugman, P.R. (1991): Increasing Returns and Economic Geography. Journal of Political Economy 99, 483-499
Kuznets, S. (1963): Quantitative Aspects of the Economic Growth of Nations. Economic Development and Cultural Change XI, 1-80
Kuznets, S. (1966): Modern Economic Growth - Rate, Structure, and Spread. London: Yale University Press
Lancaster, K. (1966): A New Approach to Consumer Theory. Journal of Political Economy 74, 132-157
Lancaster, K. (1971): Consumer Demand. New York: Columbia University Press
Lazear, E.P., Rosen, S. (1990): Male-Female Wage Differentials in Job Ladders. Journal of Labor Economics 8, S106-23
Leontief, W.W. (1936): Quantitative Input-Output Relations in the Economic System of the United States. Review of Economics and Statistics VIII, 105-25
Leontief, W.W. (1941): The Structure of American Economy, 1919-1939. New York: Oxford University Press
Lewis, W.A. (1955): The Theory of Economic Growth. London: Allen & Unwin
Lucas, R.E. (1988): On the Mechanics of Economic Development. Journal of Monetary Economics 22, 3-42
Malthus, T.R. (1933): An Essay on Population. London: J.D. Deut
Marshall, A. (1890): Principles of Economics, 1990. London: Macmillan
McCombie, J.S.L. (1988): A Synoptic View of Regional Growth and Unemployment: I – Neoclassical Theory. Urban Studies 25, 267-281
Mills, E.S., Hamilton, B.W. (1985): Urban Economics. London: Foresman and Company
Mincer, J. (1962): Labor Force Participation of Married Woman: A Study of Labor Supply, in Aspects of Labor Economics, edited by H.G. Lewis. Princeton: Princeton University Press
Miyao, T. (1987): Dynamic Urban Models. In Mills, E.S. (Ed.) Handbook of Regional and Urban Economics, Vol. II. Amsterdam: North-Holland
Mohring, H. (1961): Land Values and the Measurement of Highway Benefits. Journal of Political Economy 69, 236-249
Moomaw, R.L. (1983): Spatial Productivity Variations in Manufacturing: A Critical Survey of Cross-Sectional Analyses. International Regional Science Review 8, 1-22

Morishima, M. (1964): Equilibrium, Stability, and Growth. Oxford: Claredon Press
Morishima, M. (1969): Theory of Economic Growth. Oxford: Clarendon Press
Muth, R.F. (1961): The Spatial Structure of the Housing Market. Papers and Proceedings of the Regional Science Association 7, 207-220
Muth, R.F. (1969): Cities and Housing. Chicago: University of Chicago Press
Myrdal, G. (1957): Rich Lands and Poor. New York: Harper and Rowe
Nelson, R.R., Winter, S.G. (1982): An Evolutionary Theory of Economic Change. MA., Cambridge: Harvard University Press
Niehans, J. (1963): Economic Growth with Two Endogenous Factors. Quarterly Journal of Economics 77
Nijkamp, P., Stough, R., Verhoff, E. (1998, Eds.): Symposium "Endogenous Growth in a Regional Context." *The Annals of Regional Science* 32, No.1
Nikaido, H. (1968): Convex Structures and Economic Theory. New York: Academic Press
Oniki, H., Uzawa, H. (1965): Patterns of Trade and Investment in a Dynamic Model of International Trade. Review of Economic Studies XXXII, 15-38
Panico, C., Salvadori, N. (Eds.) (1993): Post Keynesian Theory of Growth and Distribution. Vermont: Elward Elgar Publishing Limited
Papageorgiou, Y.Y., Pines, D. (1999): An Essay on Urban Economic Theory. Boston: Kluwer Academic Publishers
Pasinetti, L.L. (1974): Growth and Income Distribution - Essays in Economic Theory. Cambridge: Cambridge University Press
Pasinetti, L.L. (1981): Structural Change and Economic Growth - A Theoretical Essay on the Dynamics of the Wealth of Nations. Cambridge: Cambridge University Press
Pasinetti, L.L. (1993): Structural Economic Dynamics - A Theory of the Economic Consequences of Human Learning. Cambridge: Cambridge University Press
Persson, I., Jonung, C. (Eds.) (1997): Economics of the Family and Family Policies. London: Routledge
Persson, I., Jonung, C. (Eds.) (1998.): Women's Work and Wages. London: Routledge
Rauch, J.E. (1991): Comparative Advantage, Geographic Advantage and the Volume of Trade. The Economic Journal 101, 1230-1244
Richardson, H.W. (1977): Regional Growth Theory. London: Macmillan
Rivera-Batiz, F. (1988): Increasing Returns, Monopolistic Competition, and Agglomeration Economies in Competition and Production. Regional Science and Urban Economics 18, 125-53
Roback, J. (1982): Wages, Rents and Quality of Life. Journal of Political Economy 90, 1257-1278
Robson, A.J. (1980): Costly Innovation and Natural Resources. International Economic Review 21, 17-30
Romer, P.M. (1986): Increasing Returns and Long-Run Growth. Journal of Political Economy 94, 1002-1037
Romer, P.M. (1987): Growth Based on Increasing Returns Due to Specialization. American Economic Review 77, 56-62

Rostow, W.W. (1960): The Stages of Economic Growth. Cambridge: Cambridge University Press

Roy, J.R., Johansson, B. (1993): A Model of Trade Flows in Differentiated Goods. The Annals of Regional Science 27, 95-115

Ruffin, R.J. (1979): Growth and the Long-Run Theory of International Capital Movements. American Economic Review 69, 832-842

Salvadori, N. (1991): Post-Keynesian Theory of Distribution in the Long Run. In Nell, E.J., Semmler, W. (Eds.): Nicholas Kaldor and Mainstream Economics - Confrontation or Convergence?. London: Macmillan

Samuelson, P.A. (1965): A Theory of Induced Innovation Along Kennedy-Weizsäcker Lines. Review of Economics and Statistics XLVII, 343-56

Sato, K. (1966): The Neoclassical Theorem and Distribution of Income and Wealth. The Review of Economic Studies XXXIII, 331-336

Sato, R., Tsutsui, S. (1984): Technical Progress, the Schumpeterian Hypothesis and Market Structure. Journal of Economics S4, 1-37

Scotchmer, S., Thisse, J.F. (1992): Space and Competition - A Puzzle. Annals of Regional Science 26, 269-286

Simon, J.L., Love, D.O. (1990): City Size, Prices, and Efficiency for Individual Goods and Services. The Annals of Regional Science 24, 163-175

Siebert, H. (1969): Regional Growth - Theory and Policy. Scranton: International Textbook Company

Sivitanidou, R., Wheaton, W.C. (1992): Wage and Rent Capitalization in the Commercial Real Estate Market. Journal of Urban Economics 31, 206-229

Smith, J. (1977): Family Labor Supply over the Life Cycle. Explorations in Economic Research 4, 205-276

Sraffa, P. (1960): Production of Commodities by Means of Commodities: Prelude to a Critique of Economic Theory. Cambridge: Cambridge University Press

Stokey, N.L. (1991): The Volume and Composition of Trade Between Rich and Poor Countries. Review of Economic Studies 58, 63-80

Suarez-Villa, L. (1993): The Dynamics of Regional Invention and Innovation: Innovative Capacity and Regional Change in the Twentieth Century. Geographical Analysis 25, 147-164

Takayama, A. (1985): Mathematical Economics, 2nd. Cambridge: Cambridge University Press

Takayama, T., Judge, G.G. (1971): Spatial and Temporal Price and Allocation Models. Amsterdam: North-Holland Publishing Company

Uzawa, H. (1965): Optimal Technical Change in an Aggregative Model of Economic Growth. International Economic Review 6: 18-31

Varga, A. (1998): University Research and Regional Innovation: A Spatial Econometric Analysis of Academic Technology Transfers. Boston: Kluwer Academic Publishers

Voith, R. (1991): Capitalization of Local and Regional Attributes into Wages and Rents - Differences Across Residential, Commercial and Mixed-Use Communities. Journal of Regional Science 31, 129-145

Von Thünen, J.H. (1826): Der Isolierte Staat in Beziehung auf Landwirtschaft und Nationalekonomie . Hamburg

Walz, U. (1999): Dynamics of Regional Integration. Berlin: Physica-Verlag

Wang, J.Y. (1990): Growth, Technology Transfer, and the Long-Run Theory of International Capital Movements. Journal of International Economics 29, 255-271
Weiss, Y., Willis, R.J. (1985): Children as Collective Goods and Divorce Settlements. Journal of Labor Economics 3, 268-292
Weizsäcker, C.C. (1966): Tentative Notes on a Two-Sector Model with Induced Technical Progress. Review of Economic Studies XXXIII, 245-51
Wigno, L., Jr. (1961): Transportation and Urban Land. Washington, DC.: Resources for the Future
Yellin, J. (1974): Urban Population Distribution, Family Income, and Social Prejudice. Journal of Urban Economics 1, 21-47
Yinger, J. (1976): Racial Prejudice and Racial Residential Segregation in an Urban Model. Journal of Urban Economics 3, 383-96
Zhang, W.B. (1989): Coexistence and Separation of Two Residential Groups - An Interactional Spatial Dynamic Approach. Geographical Analysis 21, 91-102
Zhang, W.B. (1990a): Economic Dynamics - Growth and Development. Heidelberg: Springer-Verlag
Zhang, W.B. (1990b): Brain Drain and Economic Cycles with International Migration: A Case of Minimum Wage in the Unskilled Sector. Journal of Development Economics 32, 191-203
Zhang, W.B. (1991a): Synergetic Economics. Berlin: Springer
Zhang, W.B. (1991b): Regional Dynamics with Creativity and Knowledge Diffusion. The Annals of Regional Science 25, 179-191
Zhang, W.B. (1991c): Economic Development with Creativity and Knowledge Diffusion. Socio-Spatial Dynamics 2, 19-30
Zhang, W.B. (1991d): Technological Change and Urban Growth - A Neoclassical Urban Economic Approach. Ricerche Economiche XLV, 591-608
Zhang, W.B. (1992a): A Two-Country Growth Model - Knowledge Accumulation with International Interactions. Journal of Scientific & Industrial Research 31, 187-194
Zhang, W.B. (1992b): Trade and World Economic Growth - Differences in Knowledge Utilization and Creativity. Economic Letters 39, 199-206
Zhang, W.B. (1993a): Knowledge, Economic Geography and Growth in a Two-Group Island-Economy, Technological Forecasting and Social Change 44, 405-418
Zhang, W.B. (1993b): Wages, Service Prices and Rent - Urban Division of Labor and Amenities. Seoul Journal of Economics 6, 97-113
Zhang, W.B. (1993c): Location Choice and Land Use in an Isolated State - Endogenous Capital and Knowledge Accumulation. The Annals of Regional Science 27, 23-39
Zhang, W.B. (1993d): Woman's Labor Participation and Economic Growth - Creativity, Knowledge Utilization and Family Preference. Economics Letters 42, 105-110
Zhang, W.B. (1994a): Capital, Population and Urban Patterns. Regional Science and Urban Economics 24, 273-286
Zhang, W.B. (1994b): Knowledge, Growth and Patterns of Trade. The Annals of Regional Science 28, 285-303

Zhang, W.B. (1994c): A Multi-Country Free Trade Model with Capital Accumulation. International Economic Journal 8, 53-66

Zhang, W.B. (1995a): Leisure Time, Savings and Trade Patterns - A Two-Country Growth Model. Economic Modelling 12, 425-434

Zhang, W.B. (1995b): A Two-Country Dynamic Trade Model With Multiple Groups. International Economic Journal 9, 67-80

Zhang, W.B. (1995c): An Agricultural Equilibrium Model with Two Groups. Umeå Economic Studies No. 369, University of Umeå

Zhang, W.B. (1995d): A Two-Region Trade Equilibrium Model - Each Region with Two Classes and Two Sectors Working Paper of CERUM, University of Umeå, CWP-1995:3

Zhang, W.B. (1995e): Growth, Taste Change and Sexual Division of Labor. Seoul J. of Economics 8, 395-411

Zhang, W.B. (1996): Preference, Structure and Economic Growth. Structural Change and Economic Dynamics 7, 207-221

Zhang, W.B. (1997a): Economic Geography with Division of Labor and Amenity Difference. In Chatterji, M. (Ed.): Regional Science - Perspectives for the Future. London: Macmillan Press LTD

Zhang, W.B. (1997b): A Two-Region Model With Endogenous Capital and Knowledge - Locational Amenities and Preferences. International Review of Economics and Finance 6, 1-16

Zhang, W.B. (1997c): The Complexity of Economic Evolution - Change Speeds and Time Scales, in Complexity and Self-organization in Social and Economic Systems, edited by F.K. Fang and M. Sanglier. Berlin: Springer

Zhang, W.B. (1997d): Capital, Incomes and Life Styles. Seoul Journal of Economics 10, 119-133

Zhang, W.B. (1998a): Economic Geography with Two Regions - Capital Accumulation and Economic Structure. Australian Economic Papers 35, 225-35

Zhang, W.B. (1998b): A Two-Region Growth Model - Competition, References, Resources, and Amenities. Papers in Regional Science 77, 173-188

Zhang, W.B. (1999): Capital and Knowledge - Dynamics of Economic Structures with Non-Constant Returns. Heidelberg: Springer

Zhang, W.B. (2000): A Theory of International Trade – Capital, Knowledge and Economic Structures. Heidelberg: Springer

Zhang, W.B. (2001): An Economic Theory of Cities – Spatial Models with Capital, Knowledge, and Structures. Heidelberg: Springer

Author Index

Lecture Notes in Economics and Mathematical Systems

For information about Vols. 1–429
please contact your bookseller or Springer-Verlag

Vol. 430: J. R. Daduna, I. Branco, J. M. Pinto Paixão (Eds.), Computer-Aided Transit Scheduling. XIV, 374 pages. 1995.

Vol. 431: A. Aulin, Causal and Stochastic Elements in Business Cycles. XI, 116 pages. 1996.

Vol. 432: M. Tamiz (Ed.), Multi-Objective Programming and Goal Programming. VI, 359 pages. 1996.

Vol. 433: J. Menon, Exchange Rates and Prices. XIV, 313 pages. 1996.

Vol. 434: M. W. J. Blok, Dynamic Models of the Firm. VII, 193 pages. 1996.

Vol. 435: L. Chen, Interest Rate Dynamics, Derivatives Pricing, and Risk Management. XII, 149 pages. 1996.

Vol. 436: M. Klemisch-Ahlert, Bargaining in Economic and Ethical Environments. IX, 155 pages. 1996.

Vol. 437: C. Jordan, Batching and Scheduling. IX, 178 pages. 1996.

Vol. 438: A. Villar, General Equilibrium with Increasing Returns. XIII, 164 pages. 1996.

Vol. 439: M. Zenner, Learning to Become Rational. VII, 201 pages. 1996.

Vol. 440: W. Ryll, Litigation and Settlement in a Game with Incomplete Information. VIII, 174 pages. 1996.

Vol. 441: H. Dawid, Adaptive Learning by Genetic Algorithms. IX, 166 pages.1996.

Vol. 442: L. Corchón, Theories of Imperfectly Competitive Markets. XIII, 163 pages. 1996.

Vol. 443: G. Lang, On Overlapping Generations Models with Productive Capital. X, 98 pages. 1996.

Vol. 444: S. Jørgensen, G. Zaccour (Eds.), Dynamic Competitive Analysis in Marketing. X, 285 pages. 1996.

Vol. 445: A. H. Christer, S. Osaki, L. C. Thomas (Eds.), Stochastic Modelling in Innovative Manufactoring. X, 361 pages. 1997.

Vol. 446: G. Dhaene, Encompassing. X, 160 pages. 1997.

Vol. 447: A. Artale, Rings in Auctions. X, 172 pages. 1997.

Vol. 448: G. Fandel, T. Gal (Eds.), Multiple Criteria Decision Making. XII, 678 pages. 1997.

Vol. 449: F. Fang, M. Sanglier (Eds.), Complexity and Self-Organization in Social and Economic Systems. IX, 317 pages, 1997.

Vol. 450: P. M. Pardalos, D. W. Hearn, W. W. Hager, (Eds.), Network Optimization. VIII, 485 pages, 1997.

Vol. 451: M. Salge, Rational Bubbles. Theoretical Basis, Economic Relevance, and Empirical Evidence with a Special Emphasis on the German Stock Market.IX, 265 pages. 1997.

Vol. 452: P. Gritzmann, R. Horst, E. Sachs, R. Tichatschke (Eds.), Recent Advances in Optimization. VIII, 379 pages. 1997.

Vol. 453: A. S. Tangian, J. Gruber (Eds.), Constructing Scalar-Valued Objective Functions. VIII, 298 pages. 1997.

Vol. 454: H.-M. Krolzig, Markov-Switching Vector Autoregressions. XIV, 358 pages. 1997.

Vol. 455: R. Caballero, F. Ruiz, R. E. Steuer (Eds.), Advances in Multiple Objective and Goal Programming. VIII, 391 pages. 1997.

Vol. 456: R. Conte, R. Hegselmann, P. Terna (Eds.), Simulating Social Phenomena. VIII, 536 pages. 1997.

Vol. 457: C. Hsu, Volume and the Nonlinear Dynamics of Stock Returns. VIII, 133 pages. 1998.

Vol. 458: K. Marti, P. Kall (Eds.), Stochastic Programming Methods and Technical Applications. X, 437 pages. 1998.

Vol. 459: H. K. Ryu, D. J. Slottje, Measuring Trends in U.S. Income Inequality. XI, 195 pages. 1998.

Vol. 460: B. Fleischmann, J. A. E. E. van Nunen, M. G. Speranza, P. Stähly, Advances in Distribution Logistic. XI, 535 pages. 1998.

Vol. 461: U. Schmidt, Axiomatic Utility Theory under Risk. XV, 201 pages. 1998.

Vol. 462: L. von Auer, Dynamic Preferences, Choice Mechanisms, and Welfare. XII, 226 pages. 1998.

Vol. 463: G. Abraham-Frois (Ed.), Non-Linear Dynamics and Endogenous Cycles. VI, 204 pages. 1998.

Vol. 464: A. Aulin, The Impact of Science on Economic Growth and its Cycles. IX, 204 pages. 1998.

Vol. 465: T. J. Stewart, R. C. van den Honert (Eds.), Trends in Multicriteria Decision Making. X, 448 pages. 1998.

Vol. 466: A. Sadrieh, The Alternating Double Auction Market. VII, 350 pages. 1998.

Vol. 467: H. Hennig-Schmidt, Bargaining in a Video Experiment. Determinants of Boundedly Rational Behavior. XII, 221 pages. 1999.

Vol. 468: A. Ziegler, A Game Theory Analysis of Options. XIV, 145 pages. 1999.

Vol. 469: M. P. Vogel, Environmental Kuznets Curves. XIII, 197 pages. 1999.

Vol. 470: M. Ammann, Pricing Derivative Credit Risk. XII, 228 pages. 1999.

Vol. 471: N. H. M. Wilson (Ed.), Computer-Aided Transit Scheduling. XI, 444 pages. 1999.

Vol. 472: J.-R. Tyran, Money Illusion and Strategic Complementarity as Causes of Monetary Non-Neutrality. X, 228 pages. 1999.

Vol. 473: S. Helber, Performance Analysis of Flow Lines with Non-Linear Flow of Material. IX, 280 pages. 1999.

Vol. 474: U. Schwalbe, The Core of Economies with Asymmetric Information. IX, 141 pages. 1999.

Vol. 475: L. Kaas, Dynamic Macroeconomics with Imperfect Competition. XI, 155 pages. 1999.

Vol. 476: R. Demel, Fiscal Policy, Public Debt and the Term Structure of Interest Rates. X, 279 pages. 1999.

Vol. 477: M. Théra, R. Tichatschke (Eds.), Ill-posed Variational Problems and Regularization Techniques. VIII, 274 pages. 1999.

Vol. 478: S. Hartmann, Project Scheduling under Limited Resources. XII, 221 pages. 1999.

Vol. 479: L. v. Thadden, Money, Inflation, and Capital Formation. IX, 192 pages. 1999.

Vol. 480: M. Grazia Speranza, P. Stähly (Eds.), New Trends in Distribution Logistics. X, 336 pages. 1999.

Vol. 481: V. H. Nguyen, J. J. Strodiot, P. Tossings (Eds.). Optimation. IX, 498 pages. 2000.

Vol. 482: W. B. Zhang, A Theory of International Trade. XI, 192 pages. 2000.

Vol. 483: M. Königstein, Equity, Efficiency and Evolutionary Stability in Bargaining Games with Joint Production. XII, 197 pages. 2000.

Vol. 484: D. D. Gatti, M. Gallegati, A. Kirman, Interaction and Market Structure. VI, 298 pages. 2000.

Vol. 485: A. Garnaev, Search Games and Other Applications of Game Theory. VIII, 145 pages. 2000.

Vol. 486: M. Neugart, Nonlinear Labor Market Dynamics. X, 175 pages. 2000.

Vol. 487: Y. Y. Haimes, R. E. Steuer (Eds.), Research and Practice in Multiple Criteria Decision Making. XVII, 553 pages. 2000.

Vol. 488: B. Schmolck, Ommitted Variable Tests and Dynamic Specification. X, 144 pages. 2000.

Vol. 489: T. Steger, Transitional Dynamics and Economic Growth in Developing Countries. VIII, 151 pages. 2000.

Vol. 490: S. Minner, Strategic Safety Stocks in Supply Chains. XI, 214 pages. 2000.

Vol. 491: M. Ehrgott, Multicriteria Optimization. VIII, 242 pages. 2000.

Vol. 492: T. Phan Huy, Constraint Propagation in Flexible Manufacturing. IX, 258 pages. 2000.

Vol. 493: J. Zhu, Modular Pricing of Options. X, 170 pages. 2000.

Vol. 494: D. Franzen, Design of Master Agreements for OTC Derivatives. VIII, 175 pages. 2001.

Vol. 495: I Konnov, Combined Relaxation Methods for Variational Inequalities. XI, 181 pages. 2001.

Vol. 496: P. Weiß, Unemployment in Open Economies. XII, 226 pages. 2001.

Vol. 497: J. Inkmann, Conditional Moment Estimation of Nonlinear Equation Systems. VIII, 214 pages. 2001.

Vol. 498: M. Reutter, A Macroeconomic Model of West German Unemployment. X, 125 pages. 2001.

Vol. 499: A. Casajus, Focal Points in Framed Games. XI, 131 pages. 2001.

Vol. 500: F. Nardini, Technical Progress and Economic Growth. XVII, 191 pages. 2001.

Vol. 501: M. Fleischmann, Quantitative Models for Reverse Logistics. XI, 181 pages. 2001.

Vol. 502: N. Hadjisavvas, J. E. Martínez-Legaz, J.-P. Penot (Eds.), Generalized Convexity and Generalized Monotonicity. IX, 410 pages. 2001.

Vol. 503: A. Kirman, J.-B. Zimmermann (Eds.), Economics with Heterogenous Interacting Agents. VII, 343 pages. 2001.

Vol. 504: P.-Y. Moix (Ed.),The Measurement of Market Risk. XI, 272 pages. 2001.

Vol. 505: S. Voß, J. R. Daduna (Eds.), Computer-Aided Scheduling of Public Transport. XI, 466 pages. 2001.

Vol. 506: B. P. Kellerhals, Financial Pricing Models in Continuous Time and Kalman Filtering. XIV, 247 pages. 2001.

Vol. 507: M. Koksalan, S. Zionts, Multiple Criteria Decision Making in the New Millenium. XII, 481 pages. 2001.

Vol. 508: K. Neumann, C. Schwindt, J. Zimmermann, Project Scheduling with Time Windows and Scarce Resources. XI, 335 pages. 2002.

Vol. 509: D. Hornung, Investment, R&D, and Long-Run Growth. XVI, 194 pages. 2002.

Vol. 510: A. S. Tangian, Constructing and Applying Objective Functions. XII, 582 pages. 2002.

Vol. 511: M. Külpmann, Stock Market Overreaction and Fundamental Valuation. IX, 198 pages. 2002.

Vol. 512: W.-B. Zhang, An Economic Theory of Cities.XI, 220 pages. 2002.

Vol. 513: K. Marti, Stochastic Optimization Techniques. VIII, 364 pages. 2002.

Vol. 514: S. Wang, Y. Xia, Portfolio and Asset Pricing. XII, 200 pages. 2002.

Vol. 515: G. Heisig, Planning Stability in Material Requirements Planning System. XII, 264 pages. 2002.

Vol. 516: B. Schmid, Pricing Credit Linked Financial Instruments. X, 246 pages. 2002.

Vol. 517: H. I. Meinhardt, Cooperative Decision Making in Common Pool Situations. VIII, 205 pages. 2002.

Vol. 518: S. Napel, Bilateral Bargaining. VIII, 188 pages. 2002.

Vol. 519: A. Klose, G. Speranza, L. N. Van Wassenhove (Eds.), Quantitative Approaches to Distribution Logistics and Supply Chain Management. XIII, 421 pages. 2002.

Vol. 520: B. Glaser, Efficiency versus Sustainability in Dynamic Decision Making. IX, 252 pages. 2002.

Vol. 521: R. Cowan, N. Jonard (Eds.), Heterogenous Agents, Interactions and Economic Performance. XIV, 339 pages. 2003.

Vol. 522: C. Neff, Corporate Finance, Innovation, and Strategic Competition. IX, 218 pages. 2003.

Vol. 523: W.-B. Zhang, A. Theory of Interregional Dynamics. XI, 231 pages. 2003.

Printing and Binding: Strauss GmbH, Mörlenbach